T0137927

Studies in Computational Intelligence

Volume 697

Series editor

Janusz Kacprzyk, Polish Academy of Sciences, Warsaw, Poland
e-mail: kacprzyk@ibspan.waw.pl

About this Series

The series "Studies in Computational Intelligence" (SCI) publishes new developments and advances in the various areas of computational intelligence—quickly and with a high quality. The intent is to cover the theory, applications, and design methods of computational intelligence, as embedded in the fields of engineering, computer science, physics and life sciences, as well as the methodologies behind them. The series contains monographs, lecture notes and edited volumes in computational intelligence spanning the areas of neural networks, connectionist systems, genetic algorithms, evolutionary computation, artificial intelligence, cellular automata, self-organizing systems, soft computing, fuzzy systems, and hybrid intelligent systems. Of particular value to both the contributors and the readership are the short publication timeframe and the worldwide distribution, which enable both wide and rapid dissemination of research output.

More information about this series at http://www.springer.com/series/7092

Fahed Mostafa · Tharam Dillon
Elizabeth Chang

Computational Intelligence Applications to Option Pricing, Volatility Forecasting and Value at Risk

 Springer

Fahed Mostafa
Department of Computer Science
  and Computer Engineering
La Trobe University
Bundoora, VIC
Australia

Elizabeth Chang
School of Business
University of New South Wales
Canberra, ACT
Australia

Tharam Dillon
Department of Computer Science
  and Computer Engineering
La Trobe University
Bundoora, VIC
Australia

ISSN 1860-949X   ISSN 1860-9503   (electronic)
Studies in Computational Intelligence
ISBN 978-3-319-84713-9   ISBN 978-3-319-51668-4   (eBook)
DOI 10.1007/978-3-319-51668-4

Printed on acid-free paper

This Springer imprint is published by Springer Nature
The registered company is Springer International Publishing AG
The registered company address is: Gewerbestrasse 11, 6330 Cham, Switzerland

Preface

Increasingly there are many sources of uncertainty in markets. These sources of uncertainty can have adverse effects on the evaluation of portfolio risk exposure. This uncertainty in the market variables is known as market risk which characterises the potential loss of value of an asset due to movements in market factors.

Quantitative techniques to analyse individual financial instruments and a portfolio of assets are essential for measuring market risk. Quantitative models seek to capture the trends and behaviours in the data which are then used to deduce future values. In this book, market risk is grouped into four main categories: volatility forecasting, option pricing, hedging and portfolio risk management.

When developing these quantitative methods in this book, the focus has been twofold: first, to build on the existing methodologies such as the GARCH and Black–Scholes models and second to develop approaches to overcome some of the disadvantages inherent in some of the models arising from some of the underlying assumptions which have been found to not properly reflect the behaviours inherent in the markets. For instance, a computation intelligence approach and more particularly a neural network is used to learn from data the Black–Scholes implied volatility. The implied volatility forecasts, generated from the neural net, are converted to option price using the Black–Scholes formula. The neural network option-pricing capabilities are shown to be superior to the Black–Scholes and the GARCH option-pricing model. The neural network has also shown that it is able to reproduce the implied volatility well into the future whereas the GARCH option-pricing model shows deterioration in the implied volatility with time. A new method for delta hedging using this approach is also presented. The book has been structured to provide a systematic study of the issues involved in market risk and its organisation reflects that.

Chapter 1 provides a broad introduction to some of the important concepts involved in market risk. Time series models are reviewed in Chap. 2. All financial time series models and concepts considered in this book are reviewed and explained. The weakness of each of the modelling techniques is highlighted and explained with reference to research.

Chapter 3 introduces options, existing option-pricing models and hedging.

Chapter 4 provides a review of neural networks. Then a comprehensive review is provided on the neural networks research in forecasting volatility, option pricing, hedging and value-at-risk. In this review, the strength(s) and weakness(es) of each approach are explained.

Chapter 5 outlines important problems in financial forecasting including volatility forecasting, options pricing and hedging. It provides a definition of important terms necessary to the considerations in the chapters that follow.

Volatility forecasting models are considered and evaluated in Chap. 6 including the GARCH, EGARCH and mixture density models. This is followed by the explanation of the method adopted in this book including results, discussion and evaluation.

Chapter 7 considers option-pricing models including GARCH Option-Pricing Model (GOPM), BSOPM model, implied volatility and existing neural net models. The method utilised in this book is explained, and results, discussions and evaluation are given.

Value-at-risk is considered in Chap. 8 including definitions and models.

Chapter 9 provides a recapitulation and conclusions.

The book can be used by advanced undergraduate students and graduate students in its entirety. It is also of considerable importance to practitioners in the field. We hope that you have an enjoyable and profitable time from studying the book.

Every reasonable effort has been made to acknowledge the owners of copyright material. I would be pleased to hear from any copyright owner who has been omitted or incorrectly acknowledged.

Bundoora, Australia   Fahed Mostafa
Bundoora, Australia   Tharam Dillon
Canberra, Australia   Elizabeth Chang

Contents

Chapter 1
Introduction

Technological advances such as the introduction of the internet and increase of mobile devices have allowed for instant information sharing and consumption around the world. This has led world markets to become integrated, thereby increasing trading activity on the stock exchange by local and foreign investors. The purpose of each trading activity is very much dependent on the agenda of the participant. For instance, stocks can be bought or sold for different reasons such as the readjustment of the hedge position or for simple profit realisation. This random behaviour of the investors introduces a source of uncertainty to the markets that can have adverse effects on the evaluation of portfolio risk exposure. This uncertainty in the market variables is known as market risk. By definition, market risk is the potential loss of value of an asset due to movements in market factors (Fig. 1.1).

Measuring market risk requires quantitative techniques to analyse individual financial instruments and a portfolio of assets. This quantitative measure or model captures the trends and behaviours in the data which are then used to deduce future values. In this book, market risk is grouped into four main categories: volatility forecasting, option pricing, hedging and portfolio risk management (Fig. 1.2).

1.1 Volatility Forecasting

Financial time series modelling is typically studied in terms of the asset returns rather than the asset price. The returns series contain vital information for the investor such as the portfolio position. This vital information is captured using statistical models which are then used to predict the behaviour of certain market variables moving forward. Traditionally, forecasting models such as the ARIMA model, which are used for forecasting returns, cater for many statistical aspects of the returns time series such as the conditional mean. However, financial time series have been shown to exhibit stylised facts such as volatility clustering and heteroscedasticity. These key stylised facts are not captured by such models which

© Springer International Publishing AG 2017
F. Mostafa et al., *Computational Intelligence Applications to Option Pricing,
Volatility Forecasting and Value at Risk*, Studies in Computational
Intelligence 697, DOI 10.1007/978-3-319-51668-4_1

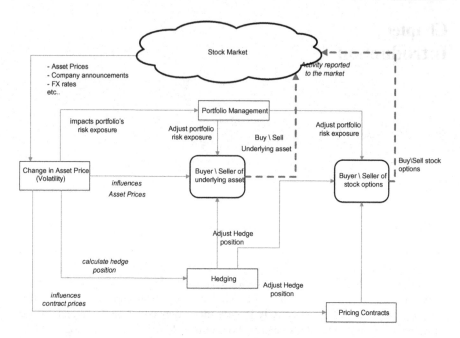

Fig. 1.1 Market risk

Fig. 1.2 Market risk categories

impacts on the reliability and accuracy of the forecasts. This problem was first addressed by the ARCH model (Engle 1982), where the variance is formulated as a function of past variances. This formulation addressed key factors in the underlying returns such as the heteroscedasticity and volatility clustering. Even though the ARCH model was a major breakthrough in volatility forecasting, the model did have several shortcomings. For this reason, many variations of the ARCH model have since emerged to address and overcome some of these shortcomings. The ARCH model family is mainly concerned with modelling the variance or volatility of the return series. The development of the ARCH model is a great contribution to market risk management as evidenced by the large body of literature that exists on this topic. Nevertheless, these forecasting models also have drawbacks which can be attributed to the prior assumptions on which they are based.

1.2   Option Pricing

Many new investment instruments have been developed and traded on the stock markets, the most common and active being options. An option is a contract on the underlying asset that can be traded depending on the investment strategy of the investor.

An option is a contract between two parties, the buyer and seller. The buyer purchases from the seller the right but not the obligation to buy or sell an asset at a fixed price in a given time frame. The buyer has to pay the seller a fee (premium) for the purchase of the option. The option that gives the buyer the right to purchase an asset is known as a call option. The option that gives the buyer the right to sell an option is known as a put option. The fixed price of the option is known as the strike price or exercise price (Fig. 1.3).

The lifetime of the option is known as the time to expiration. The option buyer has the right but not the obligation to exercise the option. If the option is not exercised, then the seller pockets the premium. If the buyer exercises the option, then the seller has to supply the underlying asset at the strike price. The two most popular option styles are the European and American options. The American style options can be exercised at any time over the life of the option, whereas European options can be exercised only at expiry. The majority of the options traded on exchanges around the world are American options. European options are generally much easier to analyse than American options. However, some of the American option properties are derived from European options.

Options are attractive instruments for investors since they can offer limited risk if applied correctly. When purchasing an option, there is a chance of gaining profit that outweighs the money spent on the premium. Typically, when an option is purchased, one does not have to exercise one's right except for the purpose of realising a profit. If the market is trading below the strike price then depending on

(a)

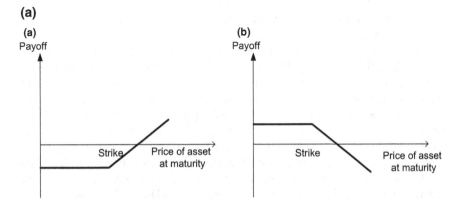

Fig. 1.3 a Profit\loss from (*a*) buying and (*b*) selling a call option. **b** Profit/loss from (*a*) buying and (*b*) selling a put option

(b)

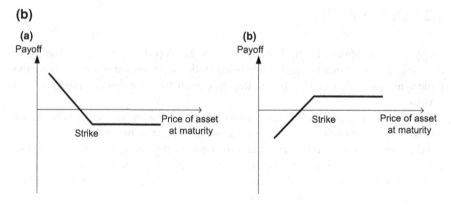

(a)

Payoff

Strike

Price of asset
at maturity

(b)

Payoff

Strike

Price of asset
at maturity

Fig. 1.3 (continued)

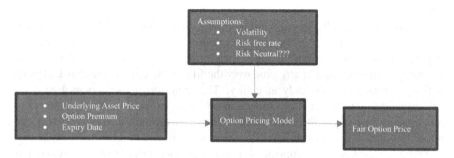

Assumptions:
- Volatility
- Risk free rate
- Risk Neutral???

- Underlying Asset Price
- Option Premium
- Expiry Date

Option Pricing Model

Fair Option Price

Fig. 1.4 Option pricing model

the type of option, an investor would make a decision about whether or not to exercise his/her right. Options also allow investors to reduce their risk exposure. Therefore, options are an essential tool in financial risk management. In this research, only European options are considered.

A mechanism for evaluating the option price is required for an effective investment strategy.

Options Pricing Models are often used for this purpose (Fig. 1.4).

The Black-Scholes Option Pricing Model (BSOPM hereafter) is the first model formulated to produce an analytical price for options. The simplicity of the model and its ability to produce a price instantaneously contributed to its wide adoption. The model was founded on several key assumptions which informed its design. These assumptions were soon found to have a significant impact on the pricing accuracy of the model. Many variations of the BSOPM have emerged to address the model's shortfalls. One critical assumption that has attracted major attention is the constant volatility assumption. This assumption is invalid as the variance in the underlying asset is time varying or heteroscedastic. This assumption is further invalidated when the volatility is extracted from the BSOPM. The implied volatility of the BSOPM is U-shaped when plotted against the moneyness of the option;

hence, the term 'volatility smile'. The implied volatility has many attractive attributes that have been widely debated in literature. The implied volatility is extracted
from the market option price; therefore, it is a forward looking view of the market.
That is, it shows the anticipation which can be exploited for further decision
making. This highlights the importance of having the correct volatility specification
in the option pricing model. To overcome the constant volatility assumption,
sophisticated models had to be developed. For instance, the GARCH option pricing
model (GOPM here after) was developed to price options by allowing the volatility
to follow a GARCH process. This feature is the key factor for superior pricing
accuracy of the GOPM. The major drawback of this model is the absence of a
closed form solution. Hence, the option pricing is generated numerically using
numeric simulation methods.

1.3   Risk Management Methods

Trading strategies are undertaken to optimise profits while controlling risk expo-
sures (Fig. 1.5).

To achieve this, one would need to understand the potential impacts on an
investment portfolio due to changes in the market variables. Hedging strategies are
designed to perform these tasks. This is achieved by using the option pricing
formula to calculate the option sensitivities to any input variable given that the
option pricing model is specified correctly. The most popular sensitivity analysis is
options delta; this measures the sensitivity of the option to one unit change in the
underlying asset price. The delta value can then be used in a hedging scenario to
offset the risk of change in the underlying asset price. On a larger scale,
Value-at-Risk models are used to provide the worst expected loss over a period of
time for a portfolio. The popularity of this model stems from its ability to

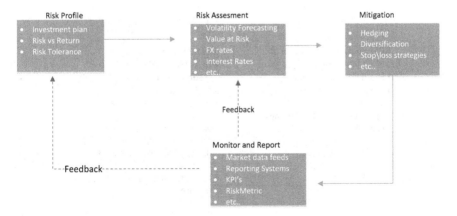

Fig. 1.5 Risk profile, assessment and mitigation

summarise the risk exposure of a portfolio into a dollar figure that can be interpreted by senior management. The wide adoption of the Value-at-Risk models is attributed to the Basle Committee making it a requirement for banks to establish minimum capital requirements based on Value-at-Risk figures. The difficulties with calculating Value-at-Risk figures is associated with capturing and predicting the extreme events. By definition, these events seldom appear in the data set; therefore, any statistical model would have difficulties representing them. This problem is generally masked by modelling the profit and loss distribution using fat tails distributions such as *student-t*. This type of assumption has a significant impact on the accuracy and performance of the model.

1.4 Neural Networks Approach

The financial models discussed above are based on statistical proprieties and assumptions in the underlying data. Assumptions are necessary for the derivation of these models. In some instances, unrealistic assumptions are made to simplify the problem or to allow the mathematical derivation of the model. Given the complex behaviour of financial markets, these models can potentially misrepresent or fail to represent key features in the underlying data. For this reason, neural networks are sought in order to overcome these shortfalls. Neural networks have been proven to be universal approximators with the capability of representing any continuous function. Neural networks are data-driven and therefore do not require prior assumptions about the data in order for the model to be formulated. In addition, they are naturally suitable to model nonlinearities in the data (Fig. 1.6).

These are very attractive features of neural networks which makes them a suitable tool for market risk modelling. These features also introduce many complexities to the design of a correct neural network model. The lack of a formal modelling approach to neural networks has resulted in many studies reporting

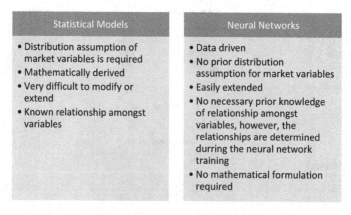

Fig. 1.6 Statistical models and neural nets

conflicting results, especially in the market risk domain. In this book, the limitation of neural networks is explained with reference to the literature. The common oversights and assumptions adopted in research and their impacts on the evaluations of the neural network models in market risk are evaluated and explained. A neural network solution is then provided for volatility forecasting, option pricing and hedging. The results achieved in this book show the superiority of neural networks over statistical models. These results highlight the important role that neural networks can play in market risk modelling and indicate that any lack of success is mainly due to the misuse and misunderstanding of neural networks.

1.5 Book Layout

The remainder of this book is organised as follows:

Time Series models are reviewed in Chap. 2. All financial time series models and concepts considered in this book are reviewed and explained. The weakness of each of the modelling techniques is highlighted and explained with reference to research. Chapter 3 introduces Options, Options Pricing Models and Hedging.

Chapter 4 provides a review of neural networks. Then a comprehensive review is provided on the neural networks research in forecasting volatility, option pricing, hedging and Value-at-Risk. In this review, the strength(s) and weakness(es) of each approach are explained.

Chapter 5 outlines important problems in Financial Forecasting including volatility forecasting, options pricing and hedging. It provides a defition of terms used in the rest of the book.

Volatility Forecasting Models are considered and evaluated in Chap. 6 including the GARCH, EGARCH and Mixture Density Models. This is followed by the explanation of the method adopted in this book including results, a discussion and evaluation.

Chapter 7 considers Options Pricing models including Garch Option Pricing Model (GOPM), the BSOPM model, Implied Volatility and existing Neural Net Models. The method utilised in this book is explained and results, discussions and evaluation are given. Value at Risk is considered in Chap. 8 including definitions and models. Chapter 9 provides a recapitulation and conclusions.

Chapter 2
Time Series Modelling

The fundamentals of time series analysis consists of a series of realisations of jointly distributed random variables, i.e. $y_1, \ldots, y_N$ The subscripts 1, ..., N are equally spaced time intervals and the observation are drawn from a probability distribution P,

$$P_{1,\ldots,N}(y_1, \ldots, y_N) \tag{2.1}$$

where $P(\cdot)$ is a probability density function associated with periods 1, ..., N with random variables $y_1, \ldots, y_N$. If the joint distribution is known at N also at time N we have observations $y_1, \ldots, y_N$ we can then construct a conditional distribution function of future observations i.e. y_{N+1}.

$$P_{N+1|1,\ldots,N}(y_{N+1}|y_1, \ldots, y_N) \tag{2.2}$$

So the information that we have about the relationship between y_1, ..., y_N and y_{N+1} from their joint distribution function allows us to specify the likely outcome of y_{N+1} using the available knowledge at N, i.e. $(y_1, \ldots, y_N)$. This model is referred to as a stochastic process, since the observation evolves through time based on the laws of probability. Random walk is a type of popular stochastic process which is used in modelling stock prices. The random walk model is a simple version of the stochastic process. The time series evolve through time according to the following:

$$y_{N+1} = y_N + \varepsilon_{N+1} \tag{2.3}$$

ε is a random variable drawn independently from a probability distribution with mean zero at every period. The probability distribution of y_{N+1} can be described given the historical observations. For instance, the mean is given by the expectation of y_{N+1} given y_1, ..., y_N,

© Springer International Publishing AG 2017
F. Mostafa et al., *Computational Intelligence Applications to Option Pricing, Volatility Forecasting and Value at Risk*, Studies in Computational Intelligence 697, DOI 10.1007/978-3-319-51668-4_2

$$E(y_{N+1}|y_1,\ldots,y_N) = E(y_N + \varepsilon_{N+1}|y_1,\ldots,y_N)$$
$$E(y_{N+1}|y_1,\ldots,y_N) = E(y_N|y_1,\ldots,y_N) + E(\varepsilon_{N+1}|y_1,\ldots,y_N)$$
$$E(y_{N+1}|y_1,\ldots,y_N) = y_N + E(\varepsilon_{N+1}) \tag{2.4}$$
$$E(y_{N+1}|y_1,\ldots,y_N) = y_N$$

Therefore, the expected value of the next evolution is the current value. Also, the variance can be calculated as follows:

$$V(y_{N+1}|y_1,\ldots,y_N) = V(y_N + \varepsilon_{N+1}|y_1,\ldots,y_N)$$
$$V(y_{N+1}|y_1,\ldots,y_N) = 0 + V(\varepsilon_{N+1}) \tag{2.5}$$
$$V(y_{N+1}|y_1,\ldots,y_N) = \sigma_\varepsilon^2$$

The forecasted value of y_{N+1} is derived from its probability distribution. For instance, if ε is assumed to be normally distributed (i.e. follows a Gaussian distribution), then we can deduce that the distribution of y_{N+1} is normally distributed centred on y_N. That is 95% of the probability enclosed in interval $y_N \pm 1.96\sigma_\varepsilon$, which means there is a 5% chance that the next observation will fall outside this interval.

2.1 Time Series Properties

2.1.1 White Noise

Consider a time series y_t which consists of independent and identically distributed (iid) random variables with finite mean and variance. If the series is normally distributed with zero mean and a variance of σ^2, the series is considered to be a Gaussian white noise. For a white noise series, all autocorrelations are zero or close to zero. In most time series modelling techniques, it is beneficial to model the serial dependence in the series beforehand, which allows the series to be de-meaned, thereby producing a white noise series. This technique is useful when modelling certain aspects of the time series such as volatility of asset returns.

2.1.2 Stochastic Processes

A stochastic process is a random process which evolves with respect to time. More precisely, the stochastic process is a collection of random variables that are indexed by time. In mathematical terms, a continuous stochastic process is defined on the probability space (Ω, F, P), where Ω is a non-empty space, F is a σ-field consisting of a subset of Ω and P is a probability measure (Billingsley 2008). This process can be described as follows: $\{x(\eta, t)\}$, where t denotes continuous time $[0, \infty)$. For a given t, $x(\eta, t)$ is a real-valued continuous random variable which maps from Ω to the real line, and η is an element of Ω (Tsay 2005).

2.1.3 Stationarity in Time Series

A stochastic process is considered to be strictly stationary if its properties are not affected by a change of time origin. That is, the distribution of r_1 is the same as other observations and the covariance between r_N and r_{N-k} is independent of N. Weak stationarity is mainly concerned with the means, variances and co-variances of the series that are independent of time rather than the entire distribution. The process r_t is defined to be weakly stationary if for all t, the following holds:

$$E[r_t] = \mu < \infty \tag{2.6}$$

$$V[r_t] = E\left[(r_t - \mu)^2\right] = \gamma_0 < \infty \tag{2.7}$$

$$\mathrm{cov}[r_t, r_{t-k}] = E[(r_t - \mu)(r_{t-k} - \mu)] = \gamma_k, \quad k = 1, 2, 3, \dots \tag{2.8}$$

Under joint normality assumption, the distribution is completely characterised by the mean and variance. So the strictly stationary and weakly stationary are equivalent. Under weak stationarity, the kth order auto covariance γ_k is,

$$\gamma_k = \mathrm{cov}[r_t, r_{t-k}] = \mathrm{cov}[r_t, r_{t+k}] \tag{2.9}$$

So when $k = 0$, γ_k gives the variance of y_t. Since the auto covariances are not independent of the units of the variables, this can be standardised by defining the autocorrelation ρ_k,

$$\rho_k = \frac{\mathrm{cov}[r_t, r_{t-k}]}{V[r_t]} = \frac{\gamma_k}{\gamma_0} \tag{2.10}$$

where $\rho_0 = 1$ and $-1 \le \rho_k \le 1$. This is also referred to as the autocorrelation function (ACF). Using the ACF, we can model the time series and the dependencies among the observations. The ACF can infer the strength and length of the process memory. It indicates how long and how strongly a shock in the process will affect the outcome. The outcome of the process could depend on the previous values of an explanatory variable. The ACF allows us to understand how to model this relationship within the time series.

2.1.4 Autoregressive Models

The autoregressive model (AR) is a regression model where the explanatory variables are lags of the dependent variable. The ACF is used to determine how many lags should be used in the AR model. This is dictated by the strength and magnitude of ρ_k. Typically, the AR model is denoted as AR(k), where k is the number of lags or past observations that are included in the regression model.

$$r_t = \alpha + \varphi\, r_{t-1} + \varepsilon_t \tag{2.11}$$

where r_{t-1} is the previous value of r and ε_t is the residual value of the process. The value of ϕ is related to the ACF and to the concept of stationarity. For $|\phi| < 1$, r_t is considered to be stationary, whereas the value of r_t will tend to keep coming back to its mean value. The time path of r_t will have no upward or downward trend, and it fluctuates around the constant mean α. When $\phi = 1$, r_t is considered to be non-stationary as it will have an upward trend. The non-stationarity implies having a unit root (i.e. $\phi = 1$). For $\phi > 1$ the process exhibits explosive behaviour over time which is uncommon in finance.

The existence of non-stationarity in the process introduces many issues when modelling time series. For instance, the persistence of shocks will be infinite for non-stationary series, which can lead to a high R^2 when the variables are not correlated. The hypothesis tests for the regression parameters cannot validly be undertaken i.e. the *t-ratios* will not follow a *t-distribution*. To overcome these sorts of issues, the time series can be differenced by subtracting r_{t-1} from both sides of Eq. 2.11,

$$\Delta r_t = \alpha + \varphi r_{t-1} + \varepsilon_t \tag{2.12}$$

where $\Delta r_t = r_t - r_{t-1}$ and $\varphi = \phi - 1$. Δr is the first difference which is usually stationary and is considered to be integrated of order 1 i.e. I(1). In some cases, it might need to be differenced twice to make it stationary (i.e. r_t and Δr_t are non-stationary, but $\Delta^2 r_t$ is stationary). In this case, r_t is said to be integrated of order 2 and written as $I(2)$.

2.2 Time Series Models

Below is a summary of time series properties and concepts used in the book.

2.2.1 The Wiener Process

A Wiener process is also known as a standard Brownian motion that is a continuous-time stochastic process with three important properties:

- It is a Markov process, where the probability distribution for all future values of the process is dependent only on the current value.
- The Wiener process has independent increments.
- Changes in the distribution are normally distributed with a variance that increases linearly with the time interval.

If $z(t)$ is a Wiener process and the change in the process is given by $\Delta z(t)$ corresponds at time interval Δt, then $z(t)$ satisfies the following:

- $\Delta z = \varepsilon_t \sqrt{\Delta t}$, where ε_t is a normally distributed random variable with zero mean and standard deviation of 1.
- $E[\varepsilon_t \varepsilon_s] = 0$ for $t \neq s$, that is ε_t is serially uncorrelated. Therefore, the values of Δz for any two different intervals of time are independent.

For a long time horizon T, the changes in the wiener process can be regarded as the sum of all the small changes up to time T. Let $N = T/\Delta t$, then z over the interval T is given by,

$$z(T) - z(0) = \sum_{i=1}^{N} \varepsilon_i \sqrt{\Delta t} \qquad (2.13)$$

ε_i (for $i = 1, ..., N$) are random draws from a normal distribution. By letting Δt become very small, the increments can be represented as a Wiener process, dz, in continuous time,

$$dz = \varepsilon_t \sqrt{dt} \qquad (2.14)$$

Since ε_t has zero mean and standard deviation of 1, the expectation and variance of the process can be defined as follows:

$$E(dz) = 0 \qquad (2.15)$$

$$V(dz) = t \qquad (2.16)$$

The Wiener process has no time derivative in the real sense; $\frac{\Delta z}{\Delta t} = \varepsilon_t \sqrt{\Delta t}$ which becomes infinite as Δt approaches zero.

2.2.2 *Geometric Brownian Motion with Drift*

The simplest generalisation of the Wiener process can be developed with a drift rate of zero and a variance of 1.

$$dx = adt + bdz \qquad (2.17)$$

where a and b are constants and dz is the Wiener process. To gain a better understanding of this relationship, let us consider each component in the right hand side of the equation adt and bdz. First let us consider the equation with the term bdz,

$$dx = adt$$
$$\text{or } \frac{dx}{dt} = a \qquad (2.18)$$

When integrating with respect to t we have,

$$x = x_0 + at \qquad (2.19)$$

where x_0 is the initial value at time 0. At time T, the value of x would have increased by aT. The bdz term can be seen as adding variability or noise to the value of x by the amount of b times to the Wiener process dz. Since the Wiener process has a standard deviation of 1, the bdz has a standard deviation of b. The change in x or δx given a small change in time δt can be defined as follows:

$$\delta x = a\delta t + b\varepsilon\sqrt{\delta t} \qquad (2.20)$$

where ε is normally distributed random variable therefore x is normally distributed with the mean $a\delta t$ and variance of $b^2\delta t$.

2.2.3 *Itô Process*

An Itô's process is a generalised Wiener process, in which parameters a and b are functions of the underlying asset variable x and time t. This can be expressed as follows:

$$dx = a(x,t)dt + b(x,t)dz \qquad (2.21)$$

The expected drift rate and the variance are able to change with time. So for a small change in time δt, the variable x changes by δx such that

$$\delta x = a(x,t)\delta t + b(x,t)\varepsilon\sqrt{\delta t} \tag{2.22}$$

This relationship assumes that the drift and variance rate of x remain constant, $a(x,t)$ and $b(x,t)^2$ respectively, during the time interval between t and δt.

2.2.4 Linear Time Series Models

The Wold decomposition theorem is typically used when modelling return series. Wold's theorem states that any de-meaned covariance stationary r_t can be represented as a sum of linearly deterministic and linearly stochastic terms:

$$r_t = \mu_t + \sum_{j=0}^{\infty} b_j\varepsilon_{t-j} \tag{2.23}$$

$$where, b_0 = 1$$

r_t is the return series, μ_t is a deterministic component or the mean, which is zero in the absence of trends in r_t. ε_t is the uncorrelated sequence which is the innovation of the process r_t. b_j is the infinite vector of moving average weights or coefficients, which is absolutely summable $\sum_{j=1}^{\infty} |b_j| < \infty$.

Let r_t be an i.i.d innovations as opposed to white noise. The unconditional mean and variance is defined as follows:

$$E[r_t] = 0 \tag{2.24}$$

$$E[r_t^2] = \sigma_\varepsilon^2 \sum_{j=0}^{\infty} b_i^2 \tag{2.25}$$

Both are invariant of time; however, the conditional mean is time varying:

$$E[r_t|\phi_{t-1}] = \sum_{i=1}^{\infty} b_i\varepsilon_{t-i} \tag{2.26}$$

where ϕ_{t-1} is the information set such that $\phi_{t-1} = \{\varepsilon_{t-1}, \varepsilon_{t-2}, \ldots\}$. Also, the conditional variance is given by

$$E[(r_t - E[r_t|\phi_{t-1}])^2|\phi_{t-1}] = \sigma_\varepsilon^2 \tag{2.27}$$

The conditional variance is a constant; therefore, it is not able to capture the dynamics of the conditional variance. The k-step-head forecast for the conditional prediction error variance is given as follows:

$$E[r_{t+k}|\phi_t] = \sum_{i=1}^{\infty} b_{k+i}\varepsilon_{t-i} \qquad (2.28)$$

$$r_{t+k} = E[r_{t+k}|\phi_t] = \sum_{i=0}^{k-1} b_i\varepsilon_{t+k-i} \qquad (2.29)$$

$$E\left[(r_{t+k} - E[r_{t+k}|\phi_t])^2\phi_t\right] = \sigma_\varepsilon^2 \sum_{i=0}^{k-1} b_i^2 \qquad (2.30)$$

As $k \to \infty$ the conditional predication error variance converges to the unconditional variance $\sigma_\varepsilon^2 \sum_{i=0}^{k-1} b_i^2$. The conditional predication error variance depends on k but not on ϕ_t. Therefore, the i.i.d innovation model does not capture the relevant information available in at time t.

2.2.5 Moving Average Model

The Moving Average (MA) model is widely used in financial time series modelling. The MA model can be expressed as an AR series.

$$r_t = \phi_0 + \phi_1 r_{t-1} + \phi_2 r_{t-2} + \cdots + \varepsilon_t \qquad (2.31)$$

The above model is unrealistic due to the infinite number of parameters. To make the model more practical, a constraint is placed on the coefficients ϕ_i so it is determined by a finite number of parameters. Let $\phi_i = -\theta_1^i$, for $i \geq 1$; then we have the following:

$$r_t = \phi_0 - \theta_1 r_{t-1} - \theta_1^2 r_{t-2} - \cdots + \varepsilon_t \qquad (2.32)$$

where $|\theta_1| < 1$ for the series is stationary. The impact of r_{t-i} decays exponentially with the increase in I ($\theta_1^i \to 0$ as $i \to \infty$). This model can be rewritten using the back-shift operator B as follows:

$$r_t = c_0 + (1 - \theta_1 B - \cdots - \theta_q B^q)\varepsilon_t \qquad (2.33)$$

where c_0 is a constant and ε_t is a white noise series.

2.2.6 Auto Regressive Moving Model (ARMA)

A natural extension of both AR and MA models would be to combine both models into an autoregressive moving average model (ARMA). This technique is popular when using volatility modelling such as the GARCH model. The ARMA (1,1) model has the following form:

$$r_t - \phi r_{t-1} = \phi_0 + \varepsilon_t - \theta_1 \varepsilon_{t-1} \tag{2.34}$$

where ε_t is a white noise series. The left hand side is the AR component and right had side is the MA part, ϕ_0 is a constant term of the equation. In the case of $\phi_1 = \theta_1$ then the process is reduced to a white noise.

Since $E[\varepsilon_t] = 0$ for all t, then given that the series is weakly stationary, the expectation of the ARMA model is given by:

$$E[r_t] = \mu = \frac{\phi_0}{1 - \phi_1} \tag{2.35}$$

with variance of

$$Var[r_t] = \frac{(1 - 2\phi_1\theta_1 + \theta_1^2)\sigma^2}{1 - \phi_1^2} \tag{2.36}$$

$\phi_1^2 < 1$ to guarantee a positive variance, which is the stationary condition for the AR(1) model. The generalised ARMA has the following form:

$$r_t = \phi_0 + \sum_{i=1}^{p} \phi_i r_{t-i} + \varepsilon_t - \sum_{i=1}^{q} \theta_i \varepsilon_{t-i} \tag{2.37}$$

2.3 Financial Time Series Modelling

Financial modelling mainly involves the modelling of the asset return series rather than asset prices. The statistical properties of the return series are more aligned to measure potential loss and profits of a portfolio which makes them more attractive than asset prices. Also, the returns series is a complete and scale-free summary of the investor's investment opportunities. Let P_t be the price of an asset at time t, then we have the following definitions for calculating the asset returns assuming the asset pays no dividends.

$$\text{Single period simple return } R_t = \frac{P_t - P_{t-1}}{P_{t-1}} \tag{2.38}$$

$$R_t(k) = \frac{(P_t - P_{t-k})}{P_{t-k}} \tag{2.39}$$

$$\text{Annualised simple return } R_t(k) \approx \frac{1}{k}\sum_{j=0}^{k-1} r_{t-j} \tag{2.40}$$

$$\text{Continuously compounded return } r_t = \ln\left(\frac{P_t}{P_{t-1}}\right) \tag{2.41}$$

2.3.1 Distributional Properties of the Return Series

Many types of distributions have been applied in modelling the distribution of the return series. The most commonly used distributions such as normal distribution, lognormal and scale mixture of normal distribution are discussed below.

2.3.1.1 Normal Distribution

It is common to assume the return series is independently and identically distributed as is normal with a constant variance and a zero mean. This assumption simplifies the modelling of the return distribution. However, the normality assumption is not supported by empirical evidence such as the existence of excess kurtosis in the returns series.

2.3.1.2 Log Normal Distribution

The log return series is generally assumed to be independently and identically distributed (iid) as normal with mean μ and variance σ^2. Therefore, the simple return is given by:

$$E(R_t) = \exp\left(\mu + \frac{\sigma^2}{2}\right) - 1, \tag{2.42}$$

$$Var(R_t) = \exp\left(2\mu + \sigma^2\right)\left[\exp(\sigma^2) - 1\right] \tag{2.43}$$

Also, let a and b be the mean and variance of the simple return that follows a lognormal distribution, then the mean and variance of the log return are given by:

$$E(r_t) = \ln\left(\frac{a+1}{\sqrt{1+b/(1+a)^2}}\right) \tag{2.44}$$

$$Var(r_t) = \ln\left(1 + \frac{b}{(1+a)^2}\right) \tag{2.45}$$

However, the log normal assumption does not exhibit some of the common features observed in the return series such as positive excess kurtosis.

2.3.1.3 Mixture of Normal Distributions

More recent studies have applied mixture of normal distributions with the attempt to capture some of the key properties of the return series distribution. Assume the log return r_t is normally distributed with mean μ and variance σ^2 or $r_t \approx N(\mu, \sigma^2)$ and that σ^2 is a random variable that follows a positive distribution such as a Gamma distribution. The mixture of normal distribution can then be written as follows:

$$r_t \sim (1-X)N(\mu, \sigma_1^2) + XN(\mu, \sigma_2^2) \tag{2.46}$$

where X is a Burnoulli random variable, for $P(X = 1) = \alpha$ and $P(X = 0) = 1 - \alpha$ with $0 < \alpha < 1$, σ_1^2 is small and σ_2^2 is relatively large. This caters for heavy tails in the distribution by increasing the value of σ_2^2. The disadvantage of this model is that it is difficult to estimate the mixture parameters.

2.3.2 Stylised Properties of Returns

It has been well documented in the literature the statistical properties of return series, where different return series from different market exhibits similar behaviour. These properties of the return series are generally referred to as stylised facts. Below is a summary of the most common stylised facts in financial return series.

- Absence of auto correlation in return series
- Returns exhibit thicker tails (i.e. leptokurtotic) than does the normal distribution.
- Gain/Loss asymmetry where the number of downward movements in the asset price is not matched by the same number of upward movements (does not apply FX-rates).

- As the time scale for the return increases, the return distribution gets closer to normal distribution. Also, the shape of the return distribution changes with the number of returns or with the time scale used.
- Intermittency of returns: at any time-scale, a high degree of variability exists in the return series.
- Volatility clustering: large changes in returns are followed by large changes and small changes tend to be followed by small changes.
- Slow decay of autocorrelation in absolute returns, which is interpreted in some cases as a sign of long-range dependence.
- Leverage effect: volatilities and asset returns are negatively correlated.
- Volume/volatility correlation: trading volume is correlated with all measures of volatility.
- Asymmetry in time scales, coarse-grained measures of volatility predict fine-scale volatility better than the other way round.
- The volatility is not constant over time: presence of heteroscedasticity in the return series.

2.3.3 Conditional Mean

Financial risk is often measured in terms of price changes to the underlying asset. This change in price can be calculated using a variety of methods, such as absolute change, log price change, relative price change etc. These price changes are known as the asset returns, which contain vital information for the investors, such as changes in portfolio value, risk exposure and investments opportunities. For this reason, they have been widely used in financial modelling. Financial models aim to capture and explain the evolution of the returns series over time. This is normally achieved by using past returns to forecast the future return values. Therefore, the model must be able to capture the underlying dynamics of the returns series as well as distribution at any given point in time.

The Wold decomposition theorem has allowed for the simplification of the time series-modelling problem. The theorem asserts that any covariance stationary process can be expressed as the sum of a deterministic and non-deterministic component.

$$P_t - P_{t-1} = \mu_t + \varepsilon_t \tag{2.47}$$

where μ_t is a deterministic process and ε_t is a non-deterministic process which are independently identically distributed (iid); this is the main concept behind the random walk model. The drawback of this formulation is the probability of obtaining a negative price in the price movements. To overcome this issue, the prices are transformed into a continuous compounded return series, by taking the log of the asset price:

$$r_t = \ln\left(\frac{P_t}{P_{t-1}}\right) \tag{2.48}$$

$$r_t = \mu_t + \varepsilon_t \tag{2.49}$$

where μ_t is mean of the return series and ε_t is *iid* with $N(0, \sigma^2)$. With this formulation, the expression for the asset price is derived as follows:

$$P_t = P_{t-1}e^{(\mu + \varepsilon_t)} \tag{2.50}$$

Therefore, P_t follows a log normal distribution.

The above random walk specification of the return series is unrealistic. The return series stylised fact exhibits many features that are contradictory to this model's specification. To capture these key features of the return series, the condition mean of the return series is typically modelled through an ARMA specification. This specification allows the return r_{t+1} to be predicted based on the previous weighted return values. The addition of the moving average component to the past disturbance also influences the future return values.

$$r_t = \phi_0 + \sum_{i=1}^{p} \phi_i r_{t-i} + \varepsilon_t - \sum_{i=1}^{q} \theta_i \varepsilon_{t-i} \tag{2.51}$$

The success of the ARMA models is well documented in the literature Tong (1990). The existence of a complete theory for linear difference equations, Gaussian models and statistical inference (with assumption of normality in ε_t) contributes to the acceptance and popularity of this modelling technique. The estimation of the model parameters can be achieved with most statistical software packages. Also, these models have been applied to many forecasting problems with different levels of success. ARMA models have been extensively adopted in finance literature; in (Poterba and Summers 1986), the Standard and Poor index (S&P) monthly returns volatility was modelled using an AR (1) process. The logarithm of the S&P monthly returns volatility in (French, Schwert et al. 1987) was modelled by a non-stationary ARIMA(0,1,3), (Schwert 1990) used an AR(12) to model a monthly volatility process. In general, the ARIMA model work well as a first order approximation in time series processes.

Financial time series models must be able to capture the behaviours of the return series in order to achieve accurate forecasts and to guarantee the stability of the model. These behaviours are well documented in the literature as the return stylised facts. The linear Gaussian models have a major limitation when trying to capture or mimic such behaviour. The ARIMA model's shortcomings are due to the inability to capture the stylised facts dynamics behaviour in the data series. The main shortcoming is the assumption of a constant variance. This assumption is a key weakness of the ARIMA model, since most financial returns series are heteroscedastic. Therefore, the variance modelling of the return series is crucial in obtaining an accurate forecast of the return series. The ARMA model assumes that the data series under study is stationary; that is, the data fluctuates around a constant mean and variance. Therefore, if any two non-stationary variables exist in the model, this would produce spurious results. This problem is treated with differencing; in some cases, the data would need to be differenced more than once to become stationary. Another drawback noted in (Tong 1990) is that if ε_t is set to be a constant for all t, then the ARIMA equation (see equation A.25 in Appendix A) becomes a deterministic linear difference equation in r_t, where r_t will have a stable limit point, and r_t always tend to unique finite constant independent of the initial value. Due to the assumption of normality, it is more appropriate to use these models where the data have a negligible probability of sudden bursts of very large amplitudes at irregular time intervals. The Gaussian assumption for the ARIMA model does not fit data that has strong asymmetry. The ARMA models are also not suitable when the data series exhibits time irreversibility (Chen and Kuan 2002). Due to these restrictions, while the ARIMA models have been successful in capturing the deterministic component of the time series, they cannot capture the time changing variance and other key stylised facts of the return time series.

2.3.4 Volatility Modelling

The volatility or variance of the time series plays a key role in the majority of finance risk models. The volatility of a return series is essential in determining the current and future asset prices. However, in order to successfully model the underlying volatility process, the model needs to cater for important stylised facts such as heteroscedasticity in the data. There exist many models which address this issue, each with its own benefits and drawbacks. To date, no complete model has been designed that caters for all aspects of the return series. Recent advances in volatility models have led to much more reliable risk models that contribute to the success of many risk management models. Given the recent instability of the world market's finance institutions, investors are always searching for models that meet their needs. In this section, we provide a review of the most common volatility models described in the literature and discuss their strengths and weaknesses.

2.3.4.1 Historical Volatility

Volatility of the return series is defined as the standard deviation of return series:

$$\sigma = \sqrt{\frac{1}{T-1}\sum_{t=1}^{T}(r_t - \mu)^2} \tag{2.52}$$

This approach is simple to calculate, but it is also dependent on the number of data points available in the sample set. Generally, the larger the data set used in calculating the volatility, the higher is the accuracy achieved. However, since volatility changes over time, how relevant is the historical data to the future forecast? The general rule is to use the most recent 90–180 days (Hull 2003). This calculation has been used in many financial models, such as the random walk models.

Another method for calculating historical volatility is the historical average method. This method assumes that the volatility distribution has a stationary mean; therefore, all variations of volatility estimates are attributed to estimation measurement error. The historical average is defined as the unweighted average of volatility observed data set:

$$\bar{\sigma} = \frac{1}{T}(\sigma_t + \sigma_{t-1} + \cdots + \sigma_1) \tag{2.53}$$

This means that the forecast can be used to compare and evaluate alternative forecast models (McMillan et al. 2000). This method still has the same drawback as the model specified in Eq. 2.23.

To rectify the issue of the sample length used to calculate the historical volatility, the simple moving average method introduces a lag length of τ that can be chosen or calculated by minimising the sample error (Poon 2005):

$$\varsigma_{t+1} = \sigma_{t+1} - \hat{\sigma}_{t+1} \tag{2.54}$$

The simple average method is defined as:

$$\hat{\sigma}_{t+1} = \frac{1}{\tau}(\sigma_t + \sigma_{t-1} + \cdots + \sigma_{t-\tau-1}) \tag{2.55}$$

This method places more emphasis on the recent observations, as it is highly probable that they will influence future observations much more than will the older observations. However, it does give all observations the same weight. As can be seen from the return series stylised facts, some return periods will have a higher influence on the future returns.

The exponentially weighted moving average (EWMA) approach places weights on the observations by having more weights on recent data and allowing the weights to decay exponentially with time. The specification of the EWMA is given as follows:

$$\hat{\sigma}_{t+1} = \lambda\sigma_t^2 + (1 - \lambda)r_t^2 \tag{2.56}$$

The value of λ is estimated by minimising the in-sample forecast error ς (Poon 2005). The RiskMetrics[TM] (Morgan 1997) approach sets the λ to 0.94 as it has been found to be the average value that minimises the one-step-ahead error variance for financial assets.

2.3.5 *Conditional Heteroscedasticity Models*

The majority of the finance models assume a constant standard deviation or homoscedasticity in the return series; however, these models are not capable of capturing some of the key stylised facts in the return series. To overcome the assumption of constant volatility, (Engle 1982) formulated the Autoregressive Conditional Heteroscedastic (ARCH) model. The basic idea behind the ARCH model is that past shocks directly impact on today's volatility. This formulation caters for the heteroscedasticity and volatility clustering which has contributed to the popularity of this model. The ARCH model in some instances could require a large number of lags which introduces many variables into the equation. This issue was overcome by (Bollerslev 1986) who generalised the ARCH model into what is now known as the Generalised Autoregressive Conditional Heteroscedastic (GARCH) model. The GARCH model implies that the unconditional variance is finite, whereas the conditional variance evolves with time; therefore, the variance is a lagged variable. The attractive features of the ARCH and GARCH models have produced many variations which have been formulated in an attempt to capture some of the key stylised facts of the return series. In this research, our main focus is on the GARCH, EGARCH and the GARCH-in-mean models.

2.3.6 *ARCH Model*

The ARCH process is defined in terms of the distribution of the residual (errors) of a linear regression model. Let us assume that the return process r_t is generated by:

$$r_t = X_t \xi + \varepsilon_t, t = 1, .., T \tag{2.57}$$

where X_t is a $k \times 1$ vector of exogenous variables, which include lagged variables, ξ is a $k \times 1$ vector of regression parameters. The ARCH model characterises the distribution of the stochastic error ε_t conditional on the set of variables of the realised values: $\psi_{t-1} = \{r_{t-1}, X_{t-1}, r_{t-2}, X_{t-2}, \ldots\}$. Engle (1982) original model assumes:

$$\varepsilon_t | \psi_{t-1} \sim N(0, \delta_t) \tag{2.58}$$

$$where \; \delta_t = \alpha_0 + \alpha_1 \varepsilon_{t-1}^2 + \cdots + \alpha_q \varepsilon_{t-q}^2 \tag{2.59}$$

with $\alpha_0 > 0$ and $\alpha_i \geq 0$, $i = 1, \ldots, q$, to ensure the conditional variance remains positive. Since $\varepsilon_{t-i} = r_{t-i} - X_{t-i}\xi, i = 1, \ldots, q$, δ_t^2 is a function of ψ_{t-1}. The non-negativity restriction is to ensure that $\delta_t^2 > 0$. The upper bound on α_i is needed to make the conditional variance stationary, $\delta_t^2 = E_{t-1}[\varepsilon_t^2]$ therefore,

$$E_{t-1}[\varepsilon_t^2] = \alpha_0 + \alpha_1 \varepsilon_{t-1}^2 + \cdots + \alpha_q \varepsilon_{t-q}^2$$

$$E[\varepsilon_t^2] = \alpha_0 + \alpha_1 E[\varepsilon_t^2] + \cdots + \alpha_q E[\varepsilon_{t-q}^2] \qquad (2.60)$$

or

$$E[\varepsilon_t^2] = \alpha_0/(1 - \alpha_1 - \cdots - \alpha_q), \quad if\,(\alpha_1 + \cdots + \alpha_q < 1)$$

In a regression model, a large shock is represented by a large deviation of r_t from its mean $X_{t-i}\xi$. In the ARCH regression model, the variance of the current error ε_{t-1}, is a function of the magnitude of the lagged errors, irrespective of their signs. So small\large errors tend to be followed by small\large errors irrespective of their signs. The order of the lag q determines the length of time for which the shock persists in conditioning the variance of subsequent errors. The larger the values of q, the longer are the episodes of volatility. This specification allows the model to capture volatility clustering and heteroscedastic properties in the return series.

2.3.7 GARCH Model

The GARCH model (Bollerslev 1986) extended the conditional variance function in Eq. 2.30. The GARCH model suggests that the conditional variance can be specified as follows:

$$\sigma_t = \alpha_0 + \alpha_1 \varepsilon_{t-1}^2 + \cdots + \alpha_q \varepsilon_{t-q} + \beta_1 \sigma_{t-1} + \cdots + \beta_p \sigma_{t-p}$$
$$\alpha_0 > 0$$
$$\alpha_i \geq 0, \quad i = 1, \ldots, q \qquad (2.61)$$
$$\beta_i \geq 0, \quad i = 1, \ldots, p$$

The inequalities are imposed to ensure that the conditional variance is positive. A GARCH process with order p and q is denoted by GARCH(p,q). By expressing (2.57) as

$$\sigma_t = \alpha_0 + \alpha(B)\varepsilon_t^2 + \beta(B)\sigma_t \qquad (2.62)$$

where $\alpha(B) = \alpha_1 B + \cdots + \alpha_q B^q$ and $\beta(B) = \beta_1 B + \cdots + \beta_p B^p$, the variables are polynomials in a backshift operator B. The GARCH model is considered to be a generalisation of an ARCH(∞) process, since the conditional variance depends linearly on all previous squared residuals.

2.3.8 GARCH-in-Mean

The GARCH models are normally used to predict the risk at a given point in time of a portfolio. Therefore, a GARCH type conditional variance model can be used to represent a time-varying risk premium in explaining the excess returns which are returns compared to a riskless asset. The excess returns is the un-forecastable difference ε_t between ex-ante and ex-post rate of returns in combination with the function of the conditional variance of the portfolio. Therefore, if r_t is the excess return at time t then,

$$r_t = \mu_t + c\sigma_t^2 + a_t \tag{2.63}$$

$$a_t = \sigma_t \varepsilon_t \tag{2.64}$$

$$\sigma_t^2 = \alpha_0 + \alpha_1 a_{t-1}^2 + \beta \sigma_{t-1}^2 \tag{2.65}$$

where μ_t and c are constants; also, c is known as the risk premium parameter. A positive c indicates that the return is positively related to its volatility. There are other formulations for the GARCH-in-mean model such as $r_t = \mu_t + c\sigma_t + a_t$ and $r_t = \mu_t + c \ln(\sigma_t^2) + a_t$

2.3.9 Exponential GARCH

The GARCH model is able to capture important features in the asset returns such as volatility clustering and heteroscedasticity. However, it is not well suited for capturing a leverage effect, due to the variance equation only catering for the magnitudes of the lagged residuals and not their signs. Exponential GARCH (EARCH) was first proposed by (Nelson 1991). The EGARCH model was formulated with the variance equation which depends on the sign and size of the lagged residual.

$$\ln(\delta_t^2) = \alpha_0 + \sum_{i=1}^{p} \beta_i \ln(\delta_{t-1}^2) + \sum_{j=1}^{q} \left(\alpha_j \left| \frac{\varepsilon_{t-j}}{\delta_{t-j}} \right| - \sqrt{\frac{2}{\pi}} + \gamma_j \frac{\varepsilon_{t-j}}{\delta_{t-j}} \right) \tag{2.66}$$

The presence of leverage effects is detected by the hypothesis that $\gamma > 0$. The impact is asymmetric when $\gamma \neq 0$. The discrete time GARCH(1,1) model converges to a continuous time diffusion model as the sampling interval gets smaller (Nelson 1991). It is also argued that the ARCH models can serve as a consistent estimator for the volatility of the true underlying diffusion process even when the model is mis-specified. That is, the difference between true volatility and ARCH estimates converges to zero in probability as the sampling interval reduced. These finding bridges the gap between the theoretical continuous time models and the financial time series which is discrete in nature.

2.3.10 Time Varying Volatility Models Literature Review

Prior to the ARCH model, several informal procedures were used to address some of the modelling issues associated with the return stylised facts. For instance, variance recursive estimates over time or moving variance was used to address the time varying volatility (Bera and Higgins 1993). The introduction of the ARCH model was a breakthrough in time series modelling. The ARCH model was the first formal model designed to capture volatility clustering and heteroscedasticity in the data. This was a major contribution to volatility modelling and largely accounts for the model's wide acceptance. Given the success of Engle's ARCH model, the focus shifted from modelling the return series to modelling the returns volatility since the return series is approximately unpredictable. However, it is generally accepted that the volatility is highly predictable (Andersen et al. 2001).

The ARCH effects are documented thoroughly in the literature. The ARCH effects are shown to exist in many financial return series, (Akgiray 1989; Schwert 1990; Engle and Mustafa 1992). In (Gallant et al. 1990) ARCH effects and conditional non-normality are observed in the NYSE value-weighted index. Hsieh (1988) observed ARCH effects in the different US dollar rates where the conditional distribution of the returns changes with time. Some researchers show that the ARCH effects diminish as the frequency of the data decreases (Diebold 1988; Baillie and Bollerslev 1989; Drost and Nijman 1993). This effect is explained by Diebold (1988), Gallant et al. (1991) as being caused by the rate or quality of the information arriving to the market in clusters, or the time between the arrival of the information and the processing of this information by market participants. Engle et al. (1990) suggest that volatility clustering is caused by information processing by the market participants. Most volatility models do not specify or define the rate of arrival of the information and how it is calculated. Volatility is usually calculated using historical data which is unrelated to future events. Most importantly, the volatility models do not cater for the rate of change of the information arrival and its impact on the volatility (Nwogugu 2006). The model assumes the rate of arrival of information and trading per unit time to be constant over the forecasting horizon.

The ARCH model formulation depicts the asset returns as being serially uncorrelated but dependent, where the dependency is a quadratic function. The parameters of the ARCH model are constrained to be positive to guarantee a positive variance. The conditional variance is formulated to depict the volatility clustering impacts on the dependent variable. A large shock will cause a large deviation from the conditional mean. The variance of the error term ε_t conditional on the values of the lagged errors is an increasing function of the lagged errors magnitude. The sign of the errors does not have an impact, since the errors' terms are squared. So, large shocks of either sign are followed by large shocks of either sign, whereas small shocks tend to be followed by small shocks of either sign. The ARCH model does suffer from major shortcomings. In the ARCH model framework, only the magnitude of the shocks has an impact on the volatility. This is caused by the square of the past innovation (Campbell and Hentschel 1992) and

(Christie 1982). In a real situation, the financial asset will react differently to positive or negative shocks. The parameters of the ARCH model are restricted, which limits the ability of the ARCH model with Gaussian innovation to capture the excess kurtosis of the return distribution. The ARCH model provides a mathematical formula for describing the behaviour of the conditional variance; however, it does not give any insight into the cause of volatility. The ARCH model is also known to over-predict volatility, because of its slow response to large isolated shocks to the return series.

Bollerslev (1986) proposed the GARCH model which is a generalisation of the ARCH model. This model has a similar formulation to the ARCH model where the model is a weighted average of the past squared residuals. The GARCH model conditional variance formulation includes lags for the squared error term as well as lags of the conditional variance as regressors for the conditional variance. The GARCH process is defined in terms of order in p and q i.e. GARCH (p, q). The p specifies the number of autoregressive lags or ARCH terms and q specifies the number of moving average of lags or GARCH terms (Engle 2001). When $q = 0$, the GARCH process is reduced to a pure ARCH(p) process. If p and q are zero, the GARCH process turns into a white noise process with ε_t. The purpose of the generalisation of ARCH to GARCH is that the GARCH can parsimoniously can represent an ARCH process with high order in p (Bera and Higgins 1993). In general, a GARCH model with low order in p and q can represent an ARCH process with high order in p.

The short-term behaviour of the volatility series is dictated by the GARCH conditional variance parameters i.e. β and α. A high value of β indicates that the previous shock will continue to have an impact on the volatility and in turn cause the volatility to persist. A large value for α specifies that the volatility is sensitive to market movement and reacts accordingly. With a low value for β and a high value for α, the volatility seems to be spiky at times. The GARCH can also be modified to cater for other stylised facts such as non-trading periods and predictable events; however, it will not be able to capture leverage effects in the return series (Bollerslev and Engle 1993). The coefficients of the volatility models are typically estimated based on a regression algorithm. These coefficients are sensitive to the period and the data set used in the model optimisation; therefore, the regression models are not well suited for the predicting of volatility because of the many variables involved, all of which change over time (Banerjee et al. 1986).

Generally, the GARCH models assume the normality in the return series innovations. This assumption fails to account for key stylised facts in the returns series. Milhøj (1987), Baillie and Bollerslev (1989) and McCurdy and Morgan (1988) display evidence of uncaptured stylised facts when normality is assumed, such as excess kurtosis and fat tails in the return series. This has led to the use of different distributions such as the student-t distribution (Bollerslev 1987), normal-Poisson mixture distribution in Jorion (1988) and power exponential distribution in Baillie and Bollerslev (1989). The failure to model fat tail properties of the return series can lead to spurious results (Baillie and DeGennaro 1990). The non-negativity of the constraints of the GARCH parameters can cause difficulties in the estimation procedure (Rabemananjara and Zakoïan 1993) and any cyclical or non-linear behaviour in volatility will be missed. Also, the conditional variance is unable to respond asymmetrically to the movements in ε_t. In the GARCH specification, the conditional variance is a function of past squared innovations, which means the sign of the past innovation does not have an impact on the volatility, only on the magnitude. This limits the GARCH model's ability to capture the leverage effects on the returns. To overcome some GARCH weaknesses, Nelson (1991) proposed the exponential GARCH (EGARCH). The EGARCH uses log conditional variance to relax the positive constraint of the model coefficient. The EGARCH specification will allow the variance to respond asymmetrically to rises and falls in the innovations. The advantage of this specification is the ability of the variance to respond more rapidly to negative and positive movements in the return series. This is an important stylised fact of many financial time series (Black 1976; Sentana 1995; Schwert 1990).

Nwogugu (2006) presents a comprehensive critique of the GARCH models, where some interesting observations are made with the regards to the error terms of the GARCH models. The critique states that the logic behind the GARCH model is that the error terms contains the unexplained characteristics of the dependent variable, such as the volatility, which makes it the best indicator of volatility. This cannot be an accurate assumption since the error is an estimate based on fixed parameters which itself is unsuitable for modelling dynamic time series such as asset prices. The errors also contain many different unexplained characteristics such as psychological states and effects which are sensitive to the coefficient of the regression model. These sensitivities are not reflected in the GARCH model specifications. The regression calculations and the error terms are derived based on a distribution assumption. Whereas, most time series do not conform to any specific distribution or mixed distribution. With the current models, the parameters contain information about the causal elements of volatility. One major issue with models such as GARCH is the assumption of a fixed relationship between the dependent and independent variables. This assumption does not hold since the relationship between the asset prices and variables changes with time and GARCH models simplifies this relation with the models limited parameters.

All ARCH type models share the same shortcomings. For instance, they require a significant amount of data points in the series for robust and stable parameters estimates. Just like any modelling technique, a more complex model with a higher number of parameters tends to better fit the data. However, it seems to perform poorly out-of-sample. The ARCH models focus on one-step-ahead variance forecast. They are not designed to produce long-term variance forecasts. When forecasting a few periods ahead, the conditional variance forecasts can no longer incorporate new information and will converge to the long run variance. Another issue which has not seen much attention in the literature is the sample size required for optimal model performance. Different authors use an ad hoc in-sample length to optimise the GARCH model. Ng and Lam (2006) argue that for a data set less than 700, two or more optimal solutions may be found with the maximum likelihood. Also, most initial values for the parameters direct parameters to the wrong optimal solution. They show that for a traditional GARCH model, 1000 data points are sufficient to obtain an optimal solution. However, for the MEM-GARCH model, 800 points provided the best fit for the model.

Poon (2005) conducted a survey on volatility forecasting models which revealed that there is no consistent approach in evaluating the models' forecast abilities. It is observed that models such as the EGARCH model that cater for asymmetry in the returns perform well. This is due to the strong negative relationship between volatility and the shocks. Lee (1991), and Cao and Tsay (1992) also argue the usefulness of the EGARCH model in modelling stock indices volatility. Poon (2005), shows that the GARCH model performs better than do traditional models such as EWMA in all sub-periods and under all evaluation measures. The study also shows a preference of exponential smoothing methods over GARCH for volatility forecasting. This is mainly due to the convergence issues with the GARCH models when the data periods are short or when there is a change in the volatility level. In addition, there is a significant amount of research where the results are not clear. This includes the use of different forecasting error statistics with different loss functions.

Andersen and Bollerslev (1998) consider the volatility to be a latent variable which is inherently unobservable and stochastically evolving through time. The asset volatility consists of intra-day and daily price variations which have to be measured over a certain period. This complicates the issue for forecasters, since the volatility is not directly observable and measured, but rather, it has to be estimated. This makes the forecasting performance and ranking of the model very difficult to ascertain. Since the true volatility cannot be determined exactly, volatility modelling and forecasting are transformed into a filtering problem, where the volatility is extracted with a degree of accuracy. This raises the question of how to evaluate the forecasting models. And, against what are they evaluated? Andersen and Bollerslev (1997, 1998) state that the failure of the GARCH models to provide a good forecast is not a failure of the GARCH model, but a failure to specify correctly the true volatility measure. They argue that the standard approach of using squared daily returns as a proxy for the true volatility for the daily forecasts is flawed. These measures consist of large and noisy independent zero mean constant variance error term which is unrelated to the true volatility. They suggest cumulative returns squared from intra-day data as an alternative for the daily squared return measure. This finding advocates the use of high frequency data for empirical evaluations.

From this literature review, it is evident that the main focus is on addressing the problem of capturing different aspects of the returns series stylised facts. This is achieved by designing the right model that captures the required behaviours. What we do not see is close attention being given to the optimisation or training of these models. For instance, the choice of training sample is done in an ad hoc manner through validation and verification. This can drastically affect the performance of the forecasting models, as different models will behave differently as the number of observations in the training data changes. Also, the same model can behave differently when applied across different time series. In this book, we will explore this behaviour of the time series models in more detail.

Chapter 3
Options and Options Pricing Models

3.1 Options

An option is a contract between two parties, the buyer and seller. The buyer purchases from the seller the right but not the obligation to buy or sell an asset at a fixed price in a given time frame. The buyer has to pay the seller a fee (premium) for the purchase of the option. The option that gives the buyer the right to purchase an asset is known as a call option; whereas, the option that gives the buyer the right to sell an option is known as a put option. The fixed price of the option is known as the strike price or exercise price. The lifetime of the option is known as the time to expiration. The option buyer has the right but not the obligation to exercise the option. If the option is not exercised, then the seller pockets the premium. If the buyer exercises the option, then the seller has to supply the underlying asset at the strike price. The two most popular option styles are European and American option. The American style options can be exercised at any time over the life of the option, whereas European options can be exercised only at expiry. The majority of the options traded on exchanges around the world are American options. European options are generally much easier to analyse than are American options. However, some of the American options properties are derived from the European options.

Options are attractive instruments for investors since they can offer limited risk if applied correctly. When purchasing an option, there is a chance of gaining profit that outweighs the money spent on the premium. Typically, when an option is purchased, one does not have to exercise one's right except for the purpose of realising a profit. So if the market is trading below the strike price then depending on the type of option, an investor would make a decision about whether or not to exercise his/her right. Options also allow investors to reduce their risk exposure. Therefore, options are an essential tool in financial risk management. In this research, only European options are considered.

© Springer International Publishing AG 2017

F. Mostafa et al., *Computational Intelligence Applications to Option Pricing, Volatility Forecasting and Value at Risk*, Studies in Computational Intelligence 697, DOI 10.1007/978-3-319-51668-4_3

3.1.1 Call Options

If a call option is bought, the purchaser has obtained the right to buy the underlying asset at the strike price at a given time. The buyer would be bullish or has an expectation that the value on the underlying asset will increase during the lifetime of the option. At expiration, if the value of the underlying asset has increased by an amount greater than the strike price. Then the buyer can exercise the option to realise a profit. This is achieved by purchasing the underlying asset from the seller at the strike price, which will be less than the market value. The buyer can then sell the underlying asset at the current market value. If the value of the underlying asset is less than the strike price at expiry, then the calls option is not exercised and is said to have expired worthless. Here, the buyer would have experienced a loss equivalent to the premium.

If a call is sold, the seller has a bearish expectation or an expectation that the value of the underlying asset will decrease at expiry. At expiry if the underlying asset value is less than the strike price, the buyer would not exercise. The seller would have profited by pocketing the premium. However, if the value of the underlying asset is greater than the strike price at expiry, the seller would experience a loss since the underlying asset must be provided to the buyer at the strike price which is less than the market value (Fig. 3.1).

3.1.2 Put Options

The purchaser of a put option would have acquired the right but not the obligation to sell the underlying asset at a set price at a specified date. The purchaser of a put option has an expectation that price of the underlying asset will decrease in value by the expiry date. At expiry, if the underlying asset value is less than the strike price, the purchaser can profit by selling the underlying asset at a higher price than the

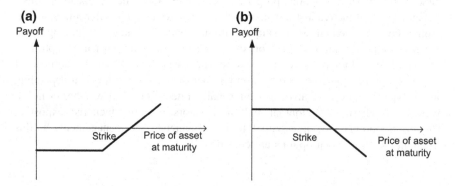

Fig. 3.1 Profit/loss from **a** buying and **b** selling a call option

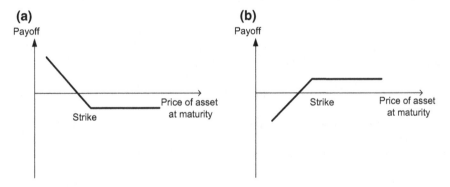

Fig. 3.2 Profit/loss from **a** buying and **b** selling a put option

market value. If the underlying asset value at expiry is higher than the strike price the purchaser will not realise any profit by exercising the option, so the put option will expire worthless.

If the put is sold, the seller would be expecting the value of the underlying asset price to fall in value such that the purchaser will not exercise the option. The seller will then realise a profit by pocketing the premium. However, if the market value of the underlying asset at expiry is greater than the strike price the option will then be exercised and the seller would have to sell the underlying asset at the strike price. The seller will then realise a loss by selling the stock at higher price than market value (Fig. 3.2).

3.1.3 Options Moneyness

The option's moneyness is determined by the value of the instrument relative to the strike price of the option. So if an option is exercised and an immediate positive cash flow is realised, that is, the stock price is greater than the strike price (for a call) then the option is said to be *in-the-money*. If the strike price of the option is equivalent to the stock price then zero cash flow will be realised, this is referred to as *at-the-money* option. If the strike price is greater than the stock price and if the option is exercised, a negative cash flow will be realised. This option is referred to as the *out-of-the-money* option.

3.1.4 Intrinsic Value

The intrinsic value of an option is the maximum of zero and the value of the option if it was exercised immediately. For a call option the intrinsic value can be specified as follows,

$$\max(K - S, 0) \tag{3.1}$$

where K is the strike price and S is the stock price. The put option intrinsic value is given by,

$$\max(0, K - S). \tag{3.2}$$

The option's intrinsic value is equivalent to the amount in money; otherwise the intrinsic value is zero.

3.1.5   Time Value

The time value of the option is derived from the time left till maturity of the option. The time value captures the possibility that the option may increase in value due to the volatility of the underlying asset. The time value can be calculated as follows,

$$\text{Time value of call} = \text{Premium} - \text{intrinsic value}.$$

The time value of an option is related to how much time it has until expiration. The more time an option has until expiration; the higher chance it has to end up in-the-money.

3.2   Option Pricing Models and Hedging

Since the publication of the Black-Scholes model in 1973 (Black and Scholes 1973), it remains the most quoted scientific paper. The Black-Scholes option pricing model (BSOPM) made a key contribution to option trading, by allowing investors to calculate a fair value of an option. This model had its limitations, which stem from the unrealistic assumptions used in deriving the model (Henderson 2004). Researchers then turned to more sophisticated methods for option valuation using stochastic volatility models to address some of the BSOPM inefficiencies, Ritchken and Trevor (1999), Peter and Kris (2004), Engle and Mustafa (1992), Duan (1995) and Heston et al. (1997). The GARCH option pricing model (GOPM) introduced by Duan (1995) is based on a discrete-time model of the economy. The GOPM is derived using the locally risk-neutralised probability measure where the option value can be calculated as a discounted expected value of the terminal asst price. In addition, the model allows the underlying asset return to follow a GARCH process (Bollerslev 1995), consequently, eliminating the assumption of a constant volatility. However, the main drawback of the model is that it does not have a closed form solution. Therefore, Monte Carlo simulation is used to optimise the models parameters. Duan and Simonato (1998) have introduced empirical martingale simulation that speeds up the Monte Carlo simulation process.

3.2.1 The Black-Scholes Options Pricing Model

The BSOPM was derived using key assumptions that allowed for the simplification and derivation of the option pricing problem. These assumptions have seen a great deal of interest, where countless research papers have been dedicated to analysing the impact of these assumptions on the accuracy and the reliability of the BSOPM. The assumptions of the BSOPM are listed below,

- **Constant Volatility**: This is the most significant assumption of the BSOPM, such that the volatility of the underlying asset remains constant over the life of the option.
- **Efficient Markets**: This suggested that market movements are unpredictable. The BSOPM monitors the stock movement using a random walk process. That is, the price of the stock at time $t + 1$ is independent of the price at time t.
- **No Dividends**: This assumption states that no dividend on the stock will be paid for the life of the option. This is an unrealistic assumption as most company's pay dividends to the shareholders. BSOPM can be adjusted for dividends by subtracting the discounted value of a future dividend from the stock price.
- **Interest rates are known and constant over the life of the option**: Just like the volatility, the interest rates are assumed to be constant over the life of the option. Typically, the government treasury bills can be used a proxy for the risk free rate. This assumption is also unrealistic as the interest rates can change over the option life time.
- **Lognormally distributed returns**: The asset return is assumed to follow a lognormal distribution. The asset return distribution generally exhibits excess kurtosis and have fatter tail then the log normal distribution.
- **European style options are used**: The BSOPM is formulated for European style options, where the option can be exercised only on the expiry date. The model has been enhanced to cater for American-style options.
- **No transaction cost or commission charges**: The BSOPM assumes no charges or fees attend the buying or selling of options.
- **Liquidity**: The Black-Scholes assumes that stocks or options can be sold or bought at any given time.

3.2.2 Black-Schole Equation

Based on the assumptions stated in Sect. 3.2.1, the BSOPM can be illustrated as follows: if $S(t)$ is an asset price that follows a Wiener process

$dS(t) = \mu S(t)dt + \sigma S(t)dz(t)$, where $z(t)$ is Brownian motion, the volatility and interest rate are constants (σ and r respectively), then a call and put option on the asset, expiring at time T with strike price K has a value at time t is given by:

$$C(t,T) = S(t)N(d_1) - Ke^{-r(T-t)}N(d_2) \tag{3.3}$$

$$P(t,T) = Ke^{-rT}N(-d_2) - S(t)N(-d_1) \tag{3.4}$$

$$d_1 = \frac{\ln\left(\frac{S(t)}{K}\right) + \left(r + \frac{\sigma^2}{\sigma}\right)(T-1)}{\sigma\sqrt{T-1}} \tag{3.5}$$

$$N(x) = (2\pi)^{-1/2} \int_{-\infty}^{x} \exp\left(-\frac{z^2}{2}\right)dz \tag{3.6}$$

$C(t,T)$ and $P(t,T)$ are the call and put price respectively. $N(x)$ is the cumulative probability distribution for a standard normally distributed variable. The Black-Scholes delta is given by $N(d_1)$. This measures the sensitivity to the underlying instrument.

3.2.3   Implied Volatility

The implied volatility is derived from the Black-Scholes model, by equating the real option price to the Black-Scholes formula and backing out the volatility parameter. The implied volatility, $\sigma_t^{BS}(K,T)$ is a function of K (Strike) and T (time to maturity) (Hull 2003).

$$C_{BS}\left(S_t, K, \tau, \sigma_t^{BS}(K,T)\right) = C_t^*(K,T) \tag{3.7}$$

where $\sigma_t^{BS}(K,T) > 0$

When the implied volatility is graphed against the strike price is takes on a U shape hence the name volatility smile (Fig. 3.3),

The two most interesting features of the implied volatility are the volatility smile (or skew) and the level of implied volatility changes with time. The volatility smile is a key indicator of unrealistic assumption of constant volatility in the BSOPM. Whereas the changes in the implied volatility level with time is seen by the deformation of the volatility surface with time. Therefore, the ability of the model to capture the deformation of the volatility surface over time will lead to accurate option pricing.

Fig. 3.3 Implied volatility

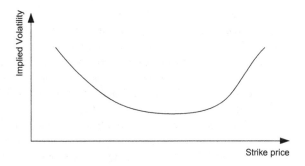

3.2.4 Black-Scholes Option Pricing Model (BSOPM)

Option pricing is an important apparatus for traders in the financial markets. The introduction of the Black Scholes model in 1973 has contributed to the growth in the option trading. The simplicity of the model allowed for traders to estimate option price instantaneously. The BSOPM have five parameters which are all observable except for the volatility parameter which needs to be estimated form the asset historical prices. The derivation of the BSOPM is made possible by the assumptions listed in Sect. 3.2.1. These assumptions are not in line with the market behaviour which raises concerns about the model accuracy.

For an in-the-money call the intrinsic value is the difference between the stock and strike price. Since the call must be priced to yield the risk free rate, today's price will equal

$$C_0 = [S_t - K](1+r)^{-T} \tag{3.8}$$

where the call is discounted at risk free rate r. Also, the stock has been priced such that it yields the risk free rate r. Therefore, the stock price is given by,

$$S_T = S_0(1+r)^T \tag{3.9}$$

then today's call price is given by,

$$C_0 = \left[S_0(1+r)^T - K\right](1+r)^{-T}$$

or

$$C_0 = S_0 - K(1+r)^{-T} \tag{3.10}$$

In the Black-Scholes world, each term is multiplied by the probabilities $N(d_1)$ and $N(d_2)$ respectively, where $N(d_2)$ is the probability that the call will end up in the money and $N(d_1)$ is the sensitivity of the call option to the underlying asset.

The popularity of the BSOPM stems from the explicit definition of an option price and the simplicity of implementing the formula to produce a fair option price. The BSOPM presents a rational framework for option pricing under general assumptions. These assumptions were soon found to be very restrictive and are in violation to the market behaviour. The assumption listed in Sect. 3.2.1, limits the model ability to capture the real market phenomena. Such assumptions were nec-essary at the time to simplify the problem which allowed for the derivation of the model. However, as the model popularity increased, researchers have been looking for ways to overcome these limitations. For instance Merton (1973) improved on the original BSOPM to incorporate dividend paying stock. Due to vast number of the research papers available to date, we will only focus on the handful of com-prehensive studies that the highlight the model pricing behaviours.

Macbeth and Merville (1979) conduct a detailed study of the BSOPM, the study revealed that the BSOPM correctly prices at-the-money options with at least 90 days-time to expiry, however the empirical analysis show biases in the BSOPM pricing. It was observed that BSOPM on average prices option less and greater than the market prices for in-the-money and out-of-the-money options respectively. With the exception of out-of-the-money options with maturity less than 90 days, the amount of under-pricing of in-the-money options and overpricing of out-of-the-money options increases with the amount to which the option is in-the-money and out-of-the-money respectively. Also, the amount of under-pricing and overpricing decreases as time to maturity gets closer to the expiry date. The BSOPM prices option with less than 90 days to maturity on average is greater than market prices. However, no consistent relationship was found between the degree of overpricing of the BSOPM and the extent of which the options are out-of-the-money or with the degree of time to expiry. These findings conflict with Black (1975) where Black stats that BSOPM generally over/under price deep in-the-money/out-of-the-money options. The results also conflict with that of Merton (1976) where he states that the BSOPM under-prices deep in-the-money options as well as deep out-of-the-money options. The authors explain that these discrepancies in the empirical research may be due to the non-stationary stochastic variance rate in the stochastic process for the generating stock prices. Another interesting point raised in this research is the fact that when the volatility is cal-culated based on historical returns the BSOPM under-prices when the BSOPM implied volatility is greater than the estimated volatility and vice versa. So one might be lead to different results when the volatility is calculated with different data over different time horizons. Hull and White (1987) conducted an empirical study of the BSOPM using stochastic volatility rather than assuming constant volatility over the life of the option. The results show that when the volatility is uncorrelated with the security prices, the BSOPM overvalues at-the-money options and under-values deep in-the-money options. However, when the volatility is positively cor-related with underlying stock price, out-of-the-money options are under-priced by the BSOPM. The reverse effect is observed when the volatility is negatively cor-related. Hammer (1989) provides a good summary of the reported BSOPM pricing biases in literature. The systematic differences between the BSOPM and market prices are reported as follows, Black (1975), Gultekin et al. (1982) and Sterk (1982) report BSOPM prices errors for in-the-money/out-of-the-money options where the BSOPM tend to be higher\ lower than market prices. In Whaley (1982) the BSOPM prices generally lower than market prices for near-to money options, also it was noticed that higher\lower the variance estimate are for the underlying asset the higher\lower the BSOPM prices tend to be relative to the market prices (Black and Scholes 1972; Whaley 1982; Gultekin et al. 1982).

The volatility is the only parameter of the BSOPM that is not directly observable in market. This parameter assumed to be constant over the life of the option. For this assumption to hold, all options on the same asset at a given date should produce the same implied volatility. In practice, different implied volatility is realised at different strike prices and time to expiry. Prior to the 1987 stock market crash, the implied volatility typical took on a U shape (i.e. volatility smile). That is options that are deep in-the-money and deep out-of-the-money exhibits higher implied volatility levels compared to the at-the-money options. After the stock market crash the implied volatility decrease monotonically as the strike price increases relative to the underlying asset price. The rate of decrease increases with short time to expiry (Dumas et al. 1998). This is mainly due to the skewed distribution of the return series to the far left, where it can be thinner than the normal distribution.

There have been several theoretical explanation for the volatility smile such as the distribution assumption and stochastic volatility (Poon 2005). There have been other explanations proposed based on market microstructure, measurement errors and investor risk preference. However, the BSOPM assumes the asset price to have a lognormal distribution or the (logarithmic) returns to follow a normal distribution. Given a high strike price the call option is deep out-of-the-money and the option has a very low probability that it will be exercised. The leptokurtic right tail will give the option a higher probability than the normal distribution for the asset price to exceed the strike price, which means the call will be in-the-money at expiry. If we consider a low strike price, the same argument applies as above but an out-of-the-money put option is used instead. Due to the thicker left tail of the leptokurtic distribution, the probability of the out-of-the-money put option to finish in-the-money will be higher than that of the normal distribution. Therefore, the put option price should be greater than the price given by the BSOPM. If the BSOPM is used to back out the implied volatility, the BSOPM implied volatility will be higher than the actual volatility. This causes the volatility smile where the implied volatility is much higher at very low strike prices and low at high strike prices. The implied volatility for different strikes and maturities do not evolve independently and they are highly correlated multivariate system (Cont and Fonseca 2002). To construct a model in terms of the implied volatility rather than volatility of the underlying asset volatility may complicate the modelling procedure. However, there are advantages of modelling the implied volatility directly. For instance, the implied volatility is observable and is derived from the market data, whereas, the asset volatility is not directly observable. The implied volatility gives an insight to the option markets which can be analysed by the practitioners. The shifts in levels of implied volatility are highly correlated across strikes and maturities, which allows for the modelling of the joint dynamics. The implied volatility is now widely used by practitioners and especially with the emergence of implied volatility indexes and derivative instruments (Cont and Fonseca 2002).

3.2.5   GARCH Option Pricing Models

The GARCH Option Pricing Model (GOPM) introduced by Duan (1995) is based on a discrete-time model of the economy. The value of the index at time t, can be assumed to have the following dynamics,

$$r_t = r_f + \lambda \sigma_t - \frac{1}{2}\sigma_t + \sigma_t \varepsilon_t \tag{3.11}$$

$$\varepsilon_t | \Omega_{t-1} \sim N(0, 1) \tag{3.12}$$

$$\sigma_t^2 = \beta_0 + \beta_1 (\varepsilon_{t-1} - \gamma)^2 \sigma_{t-1}^2 + \beta_2 \sigma_{t-1}^2 \tag{3.13}$$

Where λ is interpreted as the risk premium. To ensure the variance is positive and stationary the following constraints are applied:

$$\beta_0 > 0$$
$$\beta_1 \geq 0$$
$$\beta_2 \geq 0$$
$$\beta_1 (1 + \lambda^2) + \beta_2 < 1$$

The unconditional variance is given by $\beta_0/(1 - \beta_1(1 + \lambda^2) - \beta_2)$. If $\beta_1 = 0$ and $\beta_2 = 0$, the process is reduced to the Black-Scholes homoscedastic lognormal process when. It has been demonstrated by Duan (1995) that under the Local Risk Neutral Valuation Relationship (LRNVR) the conditional variance does not change, however under measure Q the conditional expectation of r_t is the risk free rate r_f.

$$E^Q[\exp(r_t)|\Omega_{t-1}] = \exp(r_f) \tag{3.14}$$

To derive the GOPM, the risk neutral valuation relationship has to be generalised to the LRNVR:

$$r_t = r_f - \frac{1}{2}\sigma_t^{*2} + \sigma_t^* \varepsilon_t^*,$$
$$\varepsilon_t^* \sim N(0, 1), \tag{3.15}$$

$$\sigma_t^{*2} = \beta_0 + \beta_1 (\varepsilon_{t-1}^* - \tilde{\gamma})^2 \sigma_{t-1}^{*2} + \beta_2 \sigma_{t-1}^{*2} \tag{3.16}$$

By having $\tilde{\gamma} = \lambda + \gamma$, the risk-neutral pricing measure is determined by four parameters, β_0, β_1, β_2 and $\tilde{\gamma}$. Using the above formulation, the asset terminal price is calculated at time T as follows,

$$S_T = S_t \exp\left(r_f(T-t) - \frac{1}{2}\sum_{i=t+1}^{T} \sigma_i^{*2} + \sum_{i=t+1}^{T} \sigma_i^* \varepsilon_i^* \right) \qquad (3.17)$$

The terminal asset price is obtained using Monte Carlo Simulation. A set of N random path of residuals $(\varepsilon_{t+1,j}^*, \ldots, \varepsilon_{T,j}^*)$ are generated with $J = 1$ to N. The residuals are used to calculate the asset prices $S_{T,j}$. Using the terminal asset price series the option price is then obtained by risk-neutral conditional expectation E^*:

$$C_{GARCH} = \exp(-r_f^*(T-t))E^*[\max(S_T - K, 0)] \qquad (3.18)$$

The final option price can be approximated as follow:

$$C_{GARCH} = \exp(-r_f^*(T-t))\frac{1}{N}\sum_{j=1}^{M} \max(S_{T,j} - K, 0) \qquad (3.19)$$

The delta that corresponds to the GOPM is given by Duan (1995):

$$\Delta_t^G = \exp\{-(T-t)r\}E\left[\frac{S_T}{S_t}I(S_T, K)|\varphi_t\right] \qquad (3.20)$$

Where $I(S_t, X) = 1$ is $S_t \geq K$ and 0 if $S < K$. Since there is no analytical solution for Δ_t^G the deltas are computed via Monte Carlo simulations.

Duan (1995) developed the GOPM that allows pricing of options which allows the underplaying asset returns to follow a GARCH specification. This model has key features that distinguish it from other option pricing models. In this model the option price is a function of the risk premium embedded in the underlying asset, also the underlying asset price is assumed to follow a diffusion process, thus the standard approach is of a Markovian nature. These key features can help in explaining some of the biases that exists in the BSOPM. The GOPM subsumes the BSOPM, since the heteroscedastic asset return process is a special case of the GRACH specification.

There have been several attempts to use GARCH processes to prices options, i.e. Engle and Mustafa (1992), Satchell and Timmermann (1995) and Amin and Ng (1993). Duan's GOPM extends the risk-neutralisation in Rubinstein (1976) and Brennan (1979) where the generalised version of the risk neutralisation is introduced which he called the locally risk-neutral valuation relationship (LRNVR). The LRNVR specifies the next period variance that it is not affected by changes in risk-neutral pricing measure. Therefore, the GARCH variance beyond one period is impacted by the change in measure caused by risk neutralisation. In a general sense, the difference between GOPM and BSOPM can be viewed as using different levels of volatility. Due to the GARCH process being leptokurtic, the out-of-money option is more likely to finish in the money. Which means the GOPM will be higher than the BSOPM. This also implies that the in-the-money option is more likely to finish out-of-the-money. However, this will not imply lower option prices by the GOPM when compared to BSOPM. Duan (1995) provides a numerical evaluation of the GOPM and BSOPM, in this evaluation it is noted that the BSOPM under-prices and overprices out-of-the-money option depending on the level of its initial volatility. Also the BSOPM seems to underprice deep out-of-the-money and short to maturity options. For out-of-the–money options the underpricing gets more prominent as the time to maturity is shortened. Duan concludes that the GOPM has many desirable features that can potentially correct the pricing biases of the BSOPM.

The GOPM does not have a closed form solution and the solution relies on Monte Carlo simulation which is the main drawback of this model. This is achieved by simulating different possible paths of the underlying assets, the corresponding payoff for each path is calculated at expiry then the average payoff is discounted to give the Monte Carlo price for the option. The Monte Carlo methods accuracy is inversely proportional to the square root of the number of simulation, therefore, the higher the number of simulation the more accurate the price becomes. Duan and Simonato (1998) discuss an important issue when using Monte Carlo simulation to price options. That is the occurrence of the simulated price violating rational option pricing bounds that produces a non-sensible price estimate. To overcome this issue they propose an Empirical Martingale Simulation (EMS) method which is a simple correction to the Monte Carlo simulation algorithm to ensure the simulation paths are together a martingale in an empirical sense. The advantage of EMS is that it can be coupled with other techniques such as the Antithetic and control-variate simu-lation to increase the simulation efficiency. Duan et al. (2006) follow the approach of Duan et al. (1999) and derive an analytical approximation for European option pricing that follow GJR-GARCH and EGARCH specifications. The formulation is a Black-Scholes formula with adjustments made to approximate the skewness and kurtosis of asset returns which address the simulation issues with the GOPM.

Pricing option through the GOPM is achieved through computational simula-tion. To produce an accurate price for an option a large number of simulation paths are required. This process computationally intensive and requires adequate hard-ware which can be impractical for large-scale applications. To address these issues (Duan et al. 1999) developed an analytical approximation for pricing European options in the GARCH framework. This approach is based on an Edgeworth series expansion around the normal density function. The GARCH option price is com-puted as a Black-Scholes price adjusted for skewness and kurtosis of the asset returns under a GARCH process. This method was found to be very quick to produce an option price and accurate for short maturity. It also produced accurate results for long maturity options under parameter restrictions. This method is also flexible to cater for other GARCH specifications. Their analysis shows the approximation of the GOPM produces accurate results for short maturity options.

For long maturity options, the approximation formula produces satisfactory results except when there are high levels of volatility persistence. With the high persistence the skewness and kurtosis of the returns become explosive with an increase in maturity which causes the pricing formula to perform poorly for longer-maturity options.

Heston and Nandi (2000) developed a closed-form GARCH option pricing model for European option. They provided convincing support for their model where they demonstrated its superiority over other models such as the ad hoc Black-Scholes. Their findings claim the superiority of their model is attributed to its ability to capture the correlation between volatility and returns, also capturing the volatility path dependence. The major strength of this model is existence of a closed form solution which overcomes the drawback of numerical simulations, thus this model is more practical than the GOPM. This attractive feature comes at the cost of a restrictive volatility process which can have adverse effects.

In Lehar et al. (2002), GOPM is compared with the BSOPM and the Hull-White option pricing model. The evaluations of the models are based on statistical and economic loss. The statistical errors are calculated based on the pricing accuracy of each of the models. The economic loss is based on the next-day's Value-at-Risk measure by using Monte Carlo simulation. The models were calibrated using FTSE-100 European options. This experiment established the GOPM as superior to the other models by achieving significant pricing improvements. The pricing difference between the Hull-White model and the BSOPM were very small. This finding demonstrates the importance of choosing the right volatility model for option pricing. To examine the model's performance in an economic sense, the VaR framework is utilised with Monte Carlo simulation to produce the expected profit/loss over a given horizon. The results showed that all models exhibited a weak fit to the realised profits and losses. It was also noted that the performance of the model is strongly dependent on the loss function applied. Another interesting point was that the GOPM failed to forecast the tails in the distribution of the option returns with sufficient accuracy. This is attributed to the assumption of normality in data which is problematic when applying VaR predications.

Christoffersen and Jacobs (2004) analyse different GARCH specifications for the purpose of option valuation. It is found that a heavily parameterised model has the best fit in-sample, as does the likelihood based comparison that uses the returns. The out-of-sample model evaluation favours the GARCH models with fewer parameters that contain only the volatility clustering and leverage effect. It is also shown that GARCH parameters estimated on the asset returns in a statistical sense should not be used to estimate the option price. That is because the density distribution of the asset returns is not the same as the risk-neutral density. Most volatility models are compared in terms of in-sample statistical measurements such as the likelihood value, which may not have an economical interpretation. It is suggested that in order to eliminate the shortcomings of the current models, a more radical departure from the GARCH model's assumptions is needed such as the deviation from normality in the innovation process.

Duan et al. (2006) introduces the GARCH-Jump models where the process of the returns and volatility is allowed to follow a jump diffusion process with correlated jump sizes. The disadvantages of these models are that they do not have a closed form solution and they rely on Monte Carlo simulation for pricing options. These models were then thoroughly examined in Duan et al. (2006) who extended the GOPM to incorporate non-normal innovations by introducing a new family of GARCH-jump option pricing. The model caters for non-normal innovation in the options' underlying asset and allows for stochastic volatility. The empirical analysis demonstrates the importance of incorporating jumps in the returns and volatilities which allow the model to capture the skewness and kurtosis in the underlying asset data. The empirical findings suggest that more complex models that capture more than the volatility clustering and leverage effect seem to have better performance, which contradicts the findings of Christoffersen and Jacobs (2004). The new model does not seem to require frequent calibration; the model's parameters were able to produce good option pricing up to one year beyond the parameters were estimated. The empirical evidence also shows that jumps happen more than once a day. This validates the choice of incorporating non-normal innovations and jumps in the returns and volatilities.

Hsieh and Ritchken (2005) conducted a thorough analysis of the GOPM, HN model of Heston and Nandi (2000) and the BSOPM. The empirical results show that both models are able to capture a significant amount of the strike and maturity bias in the BSOPM. The GOPM seem to outperform the HN model, especially with deep out-of-the-money options. The ability of the model to explain option pricing for a prolonged time period using old option data indicates that the model is capable of capturing key features of the option pricing process. This is apparent even when the model parameters are not estimated for years—it still performs similar to the Black-Scholes (in-sample) for at-the-money and in-the-money options. For the out-of-the-money options, the GOPM perform much better than the BSOPM. Therefore, frequent calibration of the model parameters is not required to achieve good performance relative to BSOPM.

Duan (1996) proposed an approach to use the GOPM to capture the implied volatility smile. The GOPM parameters were optimised by using the implied volatility rather than option prices. This method proved to generate good out-of-sample fit for the implied volatility and for the option prices. However, the method was demonstrated using only a couple of days of data; it is difficult to ascertain the adequacy of this technique using a limited data set. Also, this method is computationally intensive due to its reliance on Monte Carlo simulations. Aboura (2005) conducted an empirical analysis of the GOPM, using FTSE-100 European style options for five ranges of maturity to compute out-of-sample one-day-ahead option prices. The results displayed severe mis-pricing for deep out-of-the-money options which has a negative impact on the overall model performance, whereas, in-the-money options had the lowest pricing error. This is common amongst option pricing models including the BSOPM. The pricing bias of the model seems to decrease with maturity, which indicates that the GOPM is capable of pricing options with a long time to maturity. This is in line with the results demonstrating that the long-term skews are more stable over time and therefore they are more predictable. Also, the implied volatility generated by the GOPM is almost downwardly linear for long maturities. As the time gets closer to the maturity, deformation in the implied volatility starts showing for the money options. The ability of the model to capture deformation in the implied volatility through time demonstrates the model's sensitivity to regime changes in the implied volatility surface.

Lehnert (2003) argues that Gaussian models cannot adequately account for certain patterns observed in options. So by utilising the GOPM framework, GOPM is reformulated to cater for conditional leptokurtosis and skewness, allowing the GOPM to be driven by skewed generalisation error distributed innovations. The in-sample results support this method where the GOPM is observed to have added flexibility. The out-of-sample results for the DAX index show improvements over the ad hoc Black-Scholes model and Heston and Nandi model. The shape of the innovation distribution used resulted in major improvements for short-term out-of-the-money put options. Therefore, by allowing the return innovation distribution to have skewness and leptokurtosis, and the volatility to have correlation with returns, the option pricing model can better explain the shape of the implied volatility smile.

3.3 Hedging

Hedging is normally undertaken for the purpose of reducing the cash flow volatility or to minimise the probability of large losses. Using hedging techniques to minimise the impact of these risks can provide direct benefits such as reducing the chances of cash crunch, financial distress, bankruptcy and efficient capital allocation which allow the company to provide consistent returns to the investors. This is achieved by offsetting the effects of uncontrollable variables such as price movements or volatility. There are also arguments presented against hedging. For instance, it is often argued that shareholders can take on risk management initiatives themselves through diversification or through hedging. Therefore, they are not dependent on the company to manage the risk for them. The issue with this argument is that it assumes that the shareholders have the same level of understanding of the company's risk exposure as does the company's management. Also, it is cheaper for the company to hedge its risk in bulk rather than each shareholder having to manage all the company's outstanding risks. Hedging instruments have evolved over the past years, where forward contracts, futures and options were the most commonly used hedging tools. In recent years, more complex and advanced derivatives of the traditional hedging tools have been developed, such as swaps, exotic options and credit options etc. In this research, only delta hedging of European call options is undertaken.

3.3.1 Delta Hedging

Let $S(t)$ be the price of a non-dividend paying stock at time t, and X is the pay-off on derivative at its maturity T. The arbitrage-free price at some earlier time t is given by $V(t,s)$ below,

$$V(t, s) = e^{-r(T-t)}E_Q[X|S(t) = s] \qquad (3.21)$$

where Q is the risk-neutral pricing measure, r is the risk free rate, and F_t is the history of the process up to time t.

The derivative pay-off can be replicated perfectly if at all times we hold

$$\Delta(t) = \frac{\partial V}{\partial S}\Big|_{(t,S(t))} \qquad (3.22)$$

units of the share, $S(t)$ and the following amount in cash

$$C_t = V(t) - \Delta(t)S(t) \qquad (3.23)$$

This hedging process works well if the correct pricing model is used for $S(t)$ with the correct values of volatility and the risk free rate. Also, we assume that there is no transactional cost associated with trading activities and the portfolio can be rebalanced continuously. In practice, the portfolio is rebalanced in a discrete time manner, since this is more practical to avoid infinite transactional cost.

In this book, a hedge portfolio is constructed in a similar fashion to that adopted by Amilon (2003), Vahamaa (2004) Schittenkopf and Dorffner (2001). A portfolio is delta hedged with one unit short in an option and Δ units of the underlying asset, the rest is put in the bank to earn the risk free rate (or interest). That is, we buy an underpriced option (where theoretical option price is higher than market price) and sell the options that are overpriced. For underpriced options, we construct the following portfolio:

We buy option $V_0^C = C_0$ and sell index $V_0^I = -I_0\Delta_0$ and put the rest in risk-free asset, $V_0^B = I_0\Delta_0 - V_0^C$.

The portfolio is then replicated as follows:

$$V_t^I = -I_t\Delta_t \qquad (3.24)$$

$$V_0^C = C_0 \qquad (3.25)$$

$$V_t^B = \exp(rt)V_{t-1}^B + I_t(\Delta_t - \Delta_{t-1}) \qquad (3.26)$$

The value of the portfolio at time T is given by:

$$V_T = V_T^B + V_T^I + V_T^C \qquad (3.27)$$

This gives us the absolute hedging error

$$\xi = \exp(rt)|V_T| \tag{3.28}$$

where, V_t^B, V_t^I and V_t^C are the amount invested in the risk free asset, index and call option at time t.

The option pricing formula provides a fair value of the option price using some key input values such as volatility, time to expiry, risk free rate etc. The mathematical representation of the option price allows for the analysis of the option price behaviour relative to inputs variables, thereby producing important risk measures which are used to understand short term sensitivities of option premium to changes in the options' variables. Mathematically speaking, these sensitivities are expressed as partial derivatives of the option price with respect to the option variable. The option sensitivities reveal vital risk information to the investor. The option sensitivities are used to measure the option behaviour in terms of changes in an individual market parameter. These sensitivities can also be used in synthetic replication of the asset to create the options economic pay-offs by trading in the underlying asset. The option sensitivities also provide the means to define the hedging or other trading strategies.

In this research, the main focus is on the most important sensitivity measure which is the delta. The delta is the measure of the expected change in the value of the option for a given change in the underlying asset price. The delta of a call option is positive which specifies the change of the option price for an increase in the underlying asset price. The delta value of a call ranges between 0 and 1; however, a deep out-of-the-money options delta value is close to zero, since the option is not responsive to changes in the underlying asset price. Deep in-the-money call options have delta close to 1 due to the sensitivity to the underlying asset price. At-the-money call options have deltas close to 0.5. A higher delta value reveals the option price changes are closely related to the changes in the asset price, the profits and losses of which can be derived from a position assumed in the underlying asset. The delta measure also provides the probability of the option being exercised, and indicates the probability of the asset price will be above or below the strike price at expiry. This allows for the option to be replicated by dynamically adjusting the portfolio consisting of asset and cash. This process is referred to as 'delta hedging' which is a key element in the derivation of many option pricing formulas including the BSOPM. By holding the delta amount in the underlying asset, any changes in the underlying asset will offset equal and opposite changes in the value of the option portfolio.

In a perfect market, the cost of delta hedging should equal the theoretical premium of the option. In reality, the hedging process is exposed to risks that are either inherited by the pricing model or which are due to the inefficiencies in the market. The hedging process assumes that funding is readily available to purchase the delta amount of the asset as well as making short selling possible. There is a cost associated with a transaction that occurs when the portfolio is re-balanced. Hence, as the frequency of the rebalancing increases the accumulated transaction cost increases as well introducing another source of risk that can outweigh the benefits of hedging. Since the delta changes are associated with the underlying asset the transaction cost cannot be predicted in advance Leland (1985) and Boyle and Vorst (1992). Discrete rebalancing of hedged portfolio has been explored by Asay and Edelsburg (1986) and Mello and Neuhaus (1998). The valuation was based on the hedging error; however, this does represent the total risk as it does not account for heteroscedasticity in hedging errors. De Giovanni et al. (2008) compare different subordinate models for option pricing and delta hedging costs with that of the BSOPM. An evaluation of different distributional hypotheses on the hedging costs was conducted using ex-post daily prices of the S&P 500 index. Taking different exercise prices into consideration, the hedging costs are lower for the subordinated models when compared to the BSOPM. This is more evident for higher frequency hedging strategies.

In Coleman et al. (2001) the dynamic hedging performance of deterministic local volatility function are compared with implied/constant volatility. They show the hedging parameters computed from implied volatility can be significant even though its ability to calibrate the option prices with different strikes and maturities. In particular, delta hedging parameters computed using implied volatility are consistently significantly larger than the exact delta when the underlying asset follows an absolute diffusion process. The use of a local volatility function to compute the delta hedging parameters for the underlying asset that follows an absolute diffusion process produced smaller absolute average hedging errors compared with the implied volatility. Vähämaa (2004) explores the inverse relationship between volatility and movements in stock prices and its effects on delta hedging. This is achieved by adjusting the Black-Scholes delta by utilising the volatility smile to account for the inverse relationship between stock returns and changes in volatility. Empirical results using the FTSE 100 index option display major improvements for the smile-adjusted delta hedge where it consistently produces small errors relative to the performance of the Black-Scholes delta hedging scenario. The smile-adjusted delta substantially outperforms the Black-Scholes delta hedging for short term out-of-the-money and at-the-money options and becomes more evident as the hedging horizon increases.

Geyer and Schwaiger (2001) experiment with delta hedging in discrete time when the underlying asset returns follow a GARCH process. They conclude that hedging strategies based on simple delta approximation yield hedging costs on average that are close to the option prices produced by Duan's GOPM. The hedging cost variances between the strategies mainly occur with the possible different time paths. For GOPM, the risk parameter is not zero and there are strong dependencies between prices and average hedging costs which is most evident for deep out-of-the-money calls. They conclude that the GOPM pricing of out-of-the-money may be a biased reference for average hedging cost produced by discrete time delta strategies. Hsing (2006) undertakes a similar investigation in his Master's thesis.

The empirical results for the BSOPM and GOPM deltas are similar for in-the-money options. However, the BSOPM delta is higher (in absolute value) than that of the GOPM for deep out-of-the-money options and the BSOPM is lower for deep in-the-money options.

The ability to replicate the changes in the option value in the underlying asset is applicable only for small changes in asset prices and where the asset price is continuous. If a large change occurs in the asset price or discontinuities exist in the movement, the hedging process is greatly impacted upon and the option price will immediately adjust to the changes. However, the hedging cost will lag since the replication cannot occur in a sufficient amount of time, thus creating a divergence in the value of the hedge relative to the option. Boyle and Emanuel (1980) and Dotsis and Markellos (2007) demonstrate the effect of using mis-estimated volatility. It is stressed that the unknown volatility tends to aggravate the skewed nature of the hedge returns. In the case of excessive mis-estimation, it overwhelms all other variables that impact upon the returns. Asay and Edelsburg (1986) find that the impact of using the wrong volatility estimates in a hedge ratio is minimal. They argue that expecting changes in volatility levels is essential for the success of dynamic hedging.

The hedging model assumption includes constant volatility and risk free rate over the life of the option. In reality, the volatility and the risk free rate change over time. This assumption impacts upon the efficiency and accuracy of the hedge scenario. Changes in volatility levels will affect the hedging process by changing the value of the option value but not the value of the asset position. Also, the volatility levels alter the options delta which in turn will affect the hedging cost. Any changes in the risk free rate will impact upon the options delta, although the change is minor. However, a change in risk free rate affects the cost of borrowings.

Chapter 4
Neural Networks and Financial Forecasting

4.1 Neural Net Models

In the 1960s, neural networks were one of the most promising and active areas of research. Neural networks were applied to a variety of problems in the hope of achieving major breakthroughs by discovering relationships within the data. The neural networks used back then were very simplistic and consisted of a single layer network such as ADALINE (Adaptive Linear Neuron). The excitement and momentum that this field enjoyed was halted by the publication of (Minsky and Papert 1969). They demonstrated that single layer neural networks are not capable of approximating a target function that is not linearly separable, such as the XOR function. In the 1980s, neural networks were revived by the back-propagating learning algorithm in addition to the multi-layer neural networks with such publications as (Minsky and Papert 1988).

The study of artificial neural networks (ANN) was loosely motivated by the biological neural systems. However, many complexities exist in the biological neural systems that are not modelled by ANNs, and there are many features of ANNs that are inconsistent with biological systems (Mitchell 1997). The ANN-related research can be broken down into the following streams:

- Exploration of the biological system properties by utilising neural networks (computational neuroscience);
- Exploration and development of systems that are capable of learning and efficiently approximating complex functions independent of whether or not they "mirror" biological ones.

In this research, "neural network" is used within the scope of the latter type.

© Springer International Publishing AG 2017
F. Mostafa et al., *Computational Intelligence Applications to Option Pricing, Volatility Forecasting and Value at Risk*, Studies in Computational Intelligence 697, DOI 10.1007/978-3-319-51668-4_4

4.1.1 Preceptron

In its simplest form, the preceptron sums all the inputs and compares them to a
threshold value θ. If the sum of the inputs is greater than the threshold, the preceptron will fire (produce an output) such as binary value of 1 (Fig. 4.1).

This operation can be represented mathematically as follows:

Let y be a preceptron that has a function of transforming a n-dimensional input
into a binary output, $y : \Re^n \rightarrow \{0, 1\}$. The input $x = (x_1, \ldots, x_n)^T \in \Re^n$ are
weighted by $w = (w_1, \ldots, w_n)^T \in \Re^n$ then summed.

$$f(x) = \sum_{j=1}^{N} (x_j w_J) = w^T x \tag{4.1}$$

If the total value is greater than the threshold, the preceptron produces an output.
This can be represented as a step function as shown below:

$$g(x) = \begin{cases} 1, f(x) \geq \theta \\ 0, f(x) < \theta \end{cases} \tag{4.2}$$

The step function can be rewritten using an indicator function:

$$g(f(x)) = 1_{[\theta,\infty)}(a) \tag{4.3}$$

Therefore, the preceptron function can be represented as follows:

$$y(x) = g(f(x))$$
$$\text{where } y(x) = 1_{[\theta,\infty)}(w^T x) \tag{4.4}$$

The activation function of the preceptron can be represented with different activation functions, the most common of which are listed below:

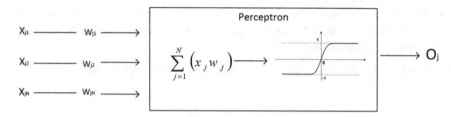

Fig. 4.1 Illustration of a Preceptron

- Linear function $\quad g(x) = ax + b$ $\hspace{4cm}$ (4.5)

- Identity function $\quad g(x) = x$ $\hspace{4cm}$ (4.6)

- Binary function $\quad g(x) = \begin{cases} 1, & \text{if } x \geq 0 \\ 0, & \text{if } x < 0 \end{cases}$ $\hspace{2cm}$ (4.7)

- Bipolar function $\quad g(x) = \begin{cases} 1, & \text{if } x \geq 0 \\ -1, & \text{if } x < 0 \end{cases}$ $\hspace{2cm}$ (4.8)

- Sigmoid function $\quad g(x) = \dfrac{1}{1 + e^{-cx}}$ $\hspace{2cm}$ (4.9)

- Hyperbolic tangent $\quad g(x) = \tanh\left(\dfrac{cx}{2}\right)$ $\hspace{2cm}$ (4.10)

4.1.2 Multi-layer Preceptron (MLP)

The common structure of a MLP is shown in Fig. 4.2. The MLP consists of three layers, input, (one or more) hidden layer and an output layer. The input layer consists of the inputs to be processed by the MLP. The hidden and output layer consists of processing units or neurons. In each layer, the units are connected to the next layer by a set of weights. The inputs are presented to the network through the input layer which is then presented for processing by each processing unit in the hidden layer. The output of each unit in the hidden layer is forwarded to the output layer to produce the final output. There is also a bias unit in each layer that is connected to each processing unit. The bias units always produce a constant output of 1 which assists in stabilising the MLP.

Since the parameters of the MLP i.e. input, weights and output are real values, we can mathematically describe the mapping function of the MLP as follows:

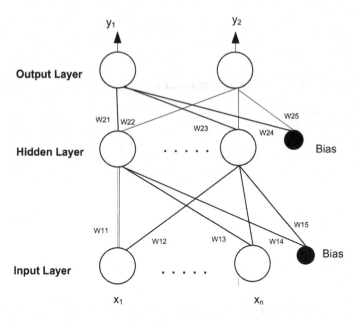

Fig. 4.2 Illustration of a typical two layer MLP

$$y_o = \psi(x, w) = f_o\big(f_h(x.w_{ih}).w_{ho}\big) \tag{4.11}$$

where y and x are the output and input of the network respectively, w is the MLP weights, ψ is the MLP function and f_o, f_h are the activation functions of the output and hidden units respectively.

The MLP maps the input space into an output space through some transformation functions or activation functions using the associated weights, where the weights are adjustable parameters in the MLP. By changing the weights, the output of the MLP changes; thus, the mapping itself changes. For the network to correctly perform the mapping between the input and outputs, the optimal weights values must be obtained to ensure minimum error between the predicted and actual values. MLP training will be discussed in more detail in Sect. 4.2.3.

MLP has been shown to have universal approximates properties, as they able to approximate any continuous function with a degree of accuracy (Haykin 2008). That is, for a single hidden layer, there exists a set of weight values that can map the input values to the output values with a given degree of accuracy. Despite this attractive feature of the MLP, unfortunately there is no formal method to determine the number of hidden units, and the correct weight values to achieve the required approximation. Even with the universal approximate property of the MLP, in some instances more than one hidden unit is required to perform complex mappings (Gibson and Cowan 1990).

4.1.3 Training MLP

The neural network training algorithm objective is to find the optimal weight values that best fit the training data. This is achieved by selecting the best set of weights that produces minimum value for an error function. Just like any optimisation problem, the error function describes how well the neural network parameters are optimised or how well the neural network is able to fit the data. The error function can be minimised by changing the weights of the neural network using such algorithms as gradient descent.

The training data set consists of a set of inputs and the corresponding output:

$$T = \{(x_1, t_1), \ldots, (x_k, t_k)\} \tag{4.12}$$

where x is the input and t is the target output. For a given input x_k, the MLP will give an output $y_k = \psi_k(T_k, w)$. Using this notation, we can derive a common error function such as the Mean Error Squared (MSE),

$$E_k = \frac{1}{2} \|y_k - \psi_k(T_k, w)\|^2 \tag{4.13}$$

$$MSE = \frac{1}{k} \sum_{i=1}^{k} E_i \tag{4.14}$$

The Sum-of-Squared error (SSE) is commonly used for training the MLPs which is a non-normalised version of the MSE

$$E = \sum_{i=1}^{k} E_i \tag{4.15}$$

After establishing the error function, we can now formulate the MLP training as an optimisation problem. Let W be the set of all feasible weights of the MLP and E is the error function that assigns a number to each feasible point $W \in \mathfrak{R}^N$, which can be expressed as:

$$E : W \rightarrow \mathfrak{R} \tag{4.16}$$

E measures the network performance and is calculated using the training data set. So to optimise the MLP to produce the desired output involves finding a set of weights w^* that produces the smallest error,

$$w^* = \arg \min_{w} E(w) \tag{4.17}$$

There are many optimisation algorithms (i.e. gradient descent) that can be used to solve this minimisation problem. There are side effects when training the MLP using SSE as the error or loss function. Thus (Bishop 1996) rewrites the SSE as follows:

$$
E^S = \frac{1}{2} \sum_{k=1}^{c} \int [f_k(x, w) - \langle t_k|x\rangle]p(x)dx
$$
$$
+ \frac{1}{2} \sum_{k=1}^{c} \int \left[\langle t_k^2|x\rangle - \langle t_k|x\rangle^2\right]p(x)dx
$$

(4.18)

The loss function is dependent on the MLP weights only in the first term, where the minimum occurs when the following minimisation criterion is met:

$$
f_k(x, w^*) = \langle t_k|x\rangle
$$

(4.19)

where w^* is the set of optimised weights that achieve the lowest SSE. Also, the conditional average of a quantity $Q(t)$ is given by:

$$
\langle Q|x\rangle \equiv \int Q(t)p(t|x)dt
$$

(4.20)

Therefore, the MLP will produce the conditional average of the target data, condition on the inputs. The second term in Eq. 2.84 is the residual value of the loss function at its global minimum, and it corresponds to the average variance of the target data around its conditional mean value. By using the SSE as a loss function for training the MLP, this allows the MLP to capture two very important statistics: the global condition mean and global variance. Bishop (1996) also derives the variance under the assumption of normality in the data to be as follows:

$$
\sigma^2 = \frac{1}{nc} \sum_{q=1}^{n} \sum_{k=1}^{c} \left[f_k(x_q, w^*) - t_k^q\right]^2
$$

(4.21)

This formulation provides the optimal value of the variance, which is given by the residual value of the SSE at its minimum with respect to the weights' optimal values w^*. The formulation also shows that the variance remains constant with respect to the inputs used. MLPs are usually trained by minimising the *SSE* function. This method is ideal for prediction of the mean $f_k(x, w^*)$, which is an approximation of the conditional average of the target data conditioned on the inputs. However, when predicting continuous variables, the conditional average represents only a small statistic which is not able to explain phenomena such as heteroscedasticity in the time series.

4.1.4 Back Propagation

The back propagation method is used to adjust the weights in the MLP to achieve the lowest error possible. The back propagation method is implemented as follows:

- Apply input x_k to the MLP and calculate the output of each unit,

$$a_j = \sum_i w_{ij} z_i \qquad (4.22)$$

$$z_j = f(a_j) \qquad (4.23)$$

where w_{ij} is the weight from unit i to j, z_i is ith input of the unit, a_j is the weight sum of inputs and z_j is the output activation of the unit activation function $f(.)$.

- Calculate the partial derivative of the error function E with respect to the weights

$$\frac{\partial E_k}{\partial w_{ji}} = \frac{\partial E_k}{\partial a_j} \cdot \frac{\partial a_j}{\partial w_{ji}} \qquad (4.24)$$

Let delta $\delta_j \equiv \frac{\partial E_k}{\partial a_j}$ and $\frac{\partial a_j}{\partial w_{ji}} = z_i$

then $\frac{\partial E_k}{\partial w_{ji}}$ can be written as $\frac{\partial E_k}{\partial w_{ji}} = \delta_j z_i$

The deltas for the output units are calculated as follows:

$$\delta_j = f'(a_j) \frac{\partial E_k}{\partial y_j} \qquad (4.25)$$

The deltas for the hidden units can also be calculated as follows:

$$\delta_j = \sum_{j=1}^{N} \frac{\partial E_k}{\partial a_j} \cdot \frac{\partial a_j}{\partial a_i} \qquad (4.26)$$

This then simplifies to

$$\delta i = f'(a_i) \sum_{j=1}^{N} w_{ji} \delta_j \qquad (4.27)$$

The gradient of the error function is given by summing all the training set:

$$\frac{\partial E}{\partial w_{ji}} = \sum_k \frac{\partial E_k}{w_{ji}} \qquad (4.28)$$

The weights are then adjusted according to the following:

$$w_{t+1} = w_t + \Delta w_t \tag{4.29}$$

$$\Delta w_t = -\eta E'(w),\ for\ \eta > 0 \tag{4.30}$$

where η is the learning rate.

The weights can be updated in two ways: firstly, they can be updated after the calculation of the gradient for the entire training set. This is referred to as batch back propagation. The second method is by on-line back propagation, which involves changing the weights for every pattern in the training set.

4.1.5 Mixture Density Networks

Mixture density networks were proposed by Bishop (1996) as an extension of the MLP to model the conditional probabilities. The MDN is constructed as an MLP; however, the output units are the parameters of a combination of Gaussian models. The proposed structure of the MDN consists of a single hidden layer with hyperbolic tan activation functions and three output units (priors, centres and widths) per Gaussian. Figure (4.3) illustrates a MDN with two Gaussians.

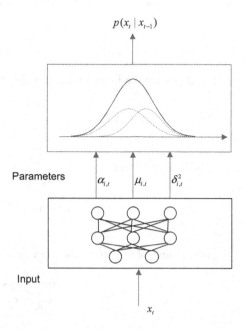

Fig. 4.3 An illustration of a mixture density network

The MDN receives the input vector x of dimensionality d and produces the conditional probability density of the target vector t of dimensionality c. A mixture of Gaussian model with M components is given by:

$$P(t|x) = \sum_{j=1}^{M} \alpha_j(x) \varphi_j(t|x) \qquad (4.31)$$

where M is the number of Gaussian components, $\alpha_j(x)$ is the mixture coefficient for the jth kernel, $\varphi_j(t|x)$ is the conditional probability of the density of the jth component. This could be any probability density function; here we only consider the Gaussian probability density function,

$$\varphi_j(t|x) = \frac{1}{(2\pi)^{\frac{c}{2}}\sigma_j} e^{-\frac{\|t-\mu(x)\|^2}{2\sigma_j(x)}} \qquad (4.32)$$

where c is the dimension of the output vector, $\mu(x)$ is the centre of the jth component. The formulation above can approximate any probability density function with an arbitrary accuracy, with the assumption that M is sufficiently large and the model parameters $\alpha_j(x)$, $\mu(x)$, and $\sigma(x)$ are chosen correctly. Therefore, it is sufficient to assume that the components of the target are statistically independent within each component of the distribution. So it is unnecessary to complicate this formulation by specifying diagonal or full covariance matrices for the variance.

The parameters of the mixture model have to satisfy some constraints, so let us denote $\theta(x)$ the number of outputs as $(c + 2)M$. The mixing coefficient $\alpha_j(x)$ must satisfy the following constraints:

$$0 \le \alpha_j(x) \le 1 \qquad (4.33)$$

$$\sum_j \alpha_j = 1 \qquad (4.34)$$

where $\alpha_j(x)$ is given by the following softmax function:

$$\alpha_j = \frac{e^{\theta_j^\alpha}}{\sum_k e^{\theta_k^\alpha}} \qquad (4.35)$$

where θ_j^α represents the corresponding networks output. The softmax function ensures the priors are positive and sum up to one. The variance is defined as

$$\sigma_j = e^{\theta_j^\sigma} \qquad (4.36)$$

θ_j^σ is the output of the MLP corresponding to the jth mixture model. The exponential function in the variance equation ensures that the variance is always positive. The centre is given by

$$\mu_{jk} = \theta_{jk}^\mu \tag{4.37}$$

θ_{jk}^μ is the output of the MLP corresponding the kth dimension of the mean vector for
the jth mixture component. Typically, when training an MLP on a non-linear
regression model, the SSE (sum squared error) is minimised.

After the MDN is optimised, it can predict the conditional density function of the
target data for any given input. This conditional density function can describe the
process generator of the data to predict the target value. For instance, the variance
produced by the MDN can be calculated as follows:

$$\sigma^2(x) = \left\langle \left\| t - \langle t|x \rangle \right\|^2 \big| x \right\rangle$$
$$= \sum_i \alpha_i(x) \left\{ \sigma_i(x)^2 + \left\| \mu(x) - \sum_j \alpha_j(x)\mu_j(x) \right\|^2 \right\} \tag{4.38}$$

This specification for the variance is a more general specification than MLP, since
the variance is a function of the input. Therefore, the variance is not constant over
the life of the dataset and can vary according to the input.

4.1.5.1   Training Mixture Density Networks

The MDN is trained by maximising the likelihood function. By assuming that the
data points are drawn independently from the same distribution, then the likelihood
function can be derived as follows:

$$L = \prod_{i=1}^{k} p(x_i, t_i) = \prod_{i=1}^{k} p(t_i|x_i)p(x) \tag{4.39}$$

L is the likelihood function of the network parameters. So the aim of training the
MDN is to maximise the likelihood function. This can also be achieved by minimising the negative log likelihood function:

$$E = -\ln L \tag{4.40}$$

$$E = -\sum_n \ln \left\{ \sum_{j=1}^{k} \alpha_j(x^n)\varphi_j(t^n|x^n) \right\} \tag{4.41}$$

The minimisation of function E with respect to the MDN parameters will allow the MDN to model the conditional density of the target. To minimise the error function E, the derivatives of the error function are calculated with respect to the weights of the neural network. If the derivative of the error function can be calculated with respect to the output values of the neural network, the back propagation method can be applied to optimise the weights. Since the error function is the sum of all errors over the training set, i.e. $E = \sum_n E^n$ then the derivative can be calculated for each pattern then summed up for all patterns.

$$\frac{\partial E_n}{\partial \theta_{jk}^\mu} = \pi_j \left\{ \frac{\mu_{jk} - t_k}{\sigma_j^2} \right\} \tag{4.42}$$

$$\frac{\partial E_n}{\partial \theta_j^\alpha} = \alpha_j - \pi_j \tag{4.43}$$

$$\frac{\partial E^n}{\partial \theta_j^\sigma} = -\pi_j \left\{ \frac{\|t - \mu_j\|^2}{\sigma_j^2} - c \right\} \tag{4.44}$$

where $\pi_j(x, t) = \frac{\alpha \phi_j}{\sum_{l=1}^{M} \alpha_l \phi_l}$, is the posterior probability which also sums up to unity, $\sum_{j=1}^{M} \pi_j = 1$. The derivatives of the error can be back propagated to calculate $\partial E / \partial w$.

Also, in the implementation, care must be taken to ensure that $\sum_{l=1}^{M} \alpha_l \phi_l \neq 0$. For a detailed explanation and derivation of MDN formulae, please refer to (Bishop 1996).

4.1.6 Radial Base Function Network

Radial Basis Function networks (RBFN) are feed-forward networks with a single hidden layer. The activation functions of the hidden units are radial basis functions. This specification is less sensitive to the non-stationary inputs and they are much faster to train compared to the MLP. The major difference between RBFN and other feedforward networks such as the MLP is the behaviour of the hidden layer. The hidden unit computes a score for the match between the input and the connecting weights. Also, the output of the RBFN is calculated by taking a linear combination of the product of the weights connecting the hidden units to the output unit.

4.1.6.1   Components of the RBFN

The RBFN mainly consists of three layers: input, hidden and output layer. The input layer consists of all inputs to the RBFN. In the hidden layer, all units are connected to all nodes in the input layer. The hidden unit has two parameters, centre and width. The number of centres c_i is the same as the number of units in the input layer. If the activation function used is a Gaussian, then the width is considered to be the variance (Fig. 4.4).

Firstly, the radial distance d_i, which is the Euclidean distance between the input vector u and the centre of the basis function c_i is required:

$$d_i = \|u - c_i\| \tag{4.45}$$

The output of the hidden unit is given by:

$$h_i = \phi(d_i, \sigma_i) \tag{4.46}$$

The output layer then performs a linear transformation to the output space. So the output is computed as follows:

$$\hat{y}(k) = w_0 + \sum_{i=1}^{M} w_{ij} \Phi_i \tag{4.47}$$

Fig. 4.4 An illustration of
the RBF network

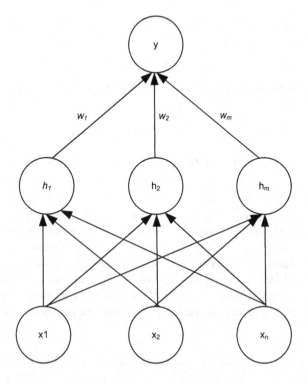

In RBF networks, the hidden and output layers play different roles when compared to the MLP network. Also, the weights are used in a complete different concept. That is why different learning algorithms are used in each layer. The weights form the input to the hidden units or the basis function parameters $\{d_i, \sigma_i\}$ can be set using unsupervised learning methods. Then the weights are kept fixed during the training of the second phase or while the hidden to output weights are learned. The second phase involves only the hidden to output weights and the output activation functions. This can be solved analytically with a set of linear equations.

4.2 Neural Networks in Financial Forecasting

In this section, a detailed literature review of neural networks research in risk management is conducted. The review highlights the strengths and weaknesses of research conducted thus far in this domain.

4.2.1 Neural Networks for Time Series Forecasting

The interest in artificial neural networks (ANNs) stems from their non-linear modelling capability. This unique learning capability, unlike conventional computing, does not require the sequence of events to be dictated in order for it to produce meaningful results. The flexibility of the neural networks allows the data to drive the model rather than forcing the model's variables to confirm to a theoretical relation that might or might not exist. This allows for the ANN to be trained without restriction of a model to discover relationships which is driven solely by the data Wallace (2008). The ANN is data-driven, self-adaptive and has few prior assumptions about the data. The ANNs learn by examples, which allows them to capture and infer key aspects and subtle relationships that may exist in the data. These key features account for the vast interest in ANNs specifically in the forecasting domain (Sharda 1994).

The generalisation ability of the ANN to infer values which are outside the population is a key feature of the ANN's modelling capability. Since forecasting involves prediction of future values from historical values, ANNs present themselves as ideal forecasting models. The ANNs are known to be universal function approximates, which is the ability to approximate any continues function with a degree of accuracy (Hornik et al. 1989, 1990; Hornik 1991); (Irie and Miyake 1998). ANNs are more flexible and adaptable form than the traditional statistical models which have limitations in estimating the relationship between historical and future values of a time series, due to the complexities in real systems. Therefore, the ANN is a good alternative to overcome this weakness.

Forecasting has been traditionally applied in the domain of linear statistics. Traditional time series methods such as the ARIMA method assume the generating process to be linear. This allows for the time series to be analysed and explained easily. However, it is inappropriate to have a prior assumption of a linear generating process, since real-world systems are usually nonlinear (Granger and Teräsvirta 1993). There has been a great deal of emphasis placed on non-linear time series models, such as threshold autoregressive (TAR) mode (Tong and Lim 1980), ARCH model (Engle 1982). These models still have major limitations, where the data series needs to be hypothesised with little knowledge of the generating process. ANNs are non-linear models which are data-driven and are able to model non-linear relationships without creating a specific hypothesis of the underlying data, therefore ANNs are more flexible models and suitable tools for forecasting (Zhang and Eddy Patuwo 1998).

According to the vast amount of literature on neural networks in finance (Medsker et al. 1993; Zhang et al. 1998) and (Wong et al. 2000), the methodology used in implementing the ANN solution differs greatly. Finding the correct model for a real problem is a cumbersome task and is equally important for ANNs as over-fitting is more likely to occur. The correct model should be able to both recognise learned patterns and generalise if an unseen pattern is presented. In Baum and Haussler (1989), a relationship was found between the generalisation capability of the ANN and size of the training set. Also Amirikian and Nishimura (1994) conclude that the size of the ANN depends on the specific tasks to be learned. Different techniques have been applied in research to construct an optimal ANN structure (Weigend et al. 1990; Weigend and Nix 1994) use a weight pruning method (weight elimination) with the addition of a term to the back-propagation cost function which penalises for the complexity of the network, (De Groot and Würtz 1991) applied a normalised Akaike information criteria (AIC) to work out the optimal ANN configuration.

Selecting the correct ANN architecture is essential in order to obtain optimal performance. The correct architecture includes selecting the correct number of hidden layers, the correct number of nodes in each of the input, hidden and output layers. To date, no formal approaches have been proposed for the selection of an optimal ANN architecture; the common approach in the literature is to apply a method of trial and error. Researchers have tried to formalise methods or test for the optimal ANN structure which are typically either heuristic or simulation-based. Examples include the polynomial time algorithm (Roy et al. 1995), the network information criteria (Murata 1994) and pruning algorithms (Sietsma and Dow 1991), the canonical decomposition technique (Wang et al. 1994). These methods are complicated to implement and do not guarantee an optimal ANN architecture for all forecasting problems. Currently, there is no universal method for determining the optimal ANN architecture.

The hidden layer plays a crucial role in the ANN modelling capabilities, where the hidden units give the ANN its generalisation capabilities and ability to perform complicated non-linear mappings. It is the hidden layer that has extended the capabilities of the ANN to model non-linear separable problems (Grossberg 1973). The number of hidden layers used in the ANN architecture is purely dependent on the problem under study. In most research, only one hidden layer is found to be sufficient for forecasting purposes. In some cases, an ANN with one hidden layer might require a large number of hidden nodes, which has a negative impact on the training time and the network generalisation ability. This issue can be addressed by adding more hidden layers which increases the complexity of the ANN architecture. In Srinivasan et al. (1994) and (Zhang et al. 1994), a two hidden layer network is found to have an advantage over a single hidden layer ANN. Other researchers use two hidden layers without comparing the results with a single hidden layer, (Vishwakarma 1994), (Grudnitski and Osburn 1993) and (Jhee and Lee 1993). Some researchers also indicate that two hidden layers are sufficient to solve most problems including forecasting problems (Lippmann 1987) and (Cybenko 1989). The optimal number of hidden nodes is also crucial to the ANN modelling capabilities. The least number of nodes is desirable which gives the ANN better generalisation capabilities and reduces the likelihood of over-fitting. There are some methods that outline different approaches to selecting the optimal number of hidden units, such as pruning unnecessary hidden units to improve ANN performance (Gorr et al. 1994). However, there is still no method that explicitly calculates the required number of hidden units. For a single hidden layer, there are several methods proposed for a number of hidden units, for example "2n + 1" (Lippmann 1987), "2n" (Wong 1991), "n/2" (Kang 1992), where n is the number of input nodes. These methods can be adopted in certain situations, but there is no method that works for all problems. Tang and Fishwick (1993) conclude that the number of hidden units have an impact on the forecasting performance of the ANN. In (De Groot and Würtz 1991); (Chakraborty et al. 1992) and (Tang and Fishwick 1993), the ANNs with an equal number of hidden units and input nodes are shown to have better forecasting performance than other configurations.

Selecting the right activation function should be done according to the problem at hand. It has been shown in Klimasauskas (1991) that the logistic activation functions are best suited to classification problems and hyperbolic tangent functions are better suited to forecasting problems. In most cases, the ANN can have the same or different activation functions for different nodes in the same or different layers (Schöneburg 1990). However, the most common practice is to use the same activation function in the same layer. The logistic function is the most common activation function used in the hidden layer. In forecasting problems where the target values are continuous in nature, a linear activation function is suitable (Chauvin and Rumelhart 1995) and (Gorr et al. 1994). In Cottrell et al. (1995), the linear activation functions in the output node cannot model times series with trend; therefore, the time series needs to be treated using methods such as differencing etc.

Another key aspect when determining the correct ANN structure for a forecasting problem is selecting the correct input. The input variables must have enough information to allow the ANN to learn the underlying process. Ideally, the least number of inputs is desirable, so that the essential features of the data can be captured in the ANN model. In the absence of a formal method for selecting the correct inputs for the ANN, some researchers try to use a standard time series approach for selecting the right input values. In Tang and Fishwick (1993), the Box-Jenkins method was used to work out the number of lagged variables into the

ANN model. This method is not ideal since the Box-Jenkins methods are for linear models and ANNs are nonlinear. The AR terms specify only the lagged terms which is not suitable for non-linear models (Zhang et al. 1998). In the literature, there are conflicting methods for determining the inputs for ANN models. Tang and Fishwick (1993) discuss the benefit of using more inputs, whereas (Lachtermacher and Fuller 1995) argue the negative impact of having more inputs on one-step-ahead forecasts, but positive results for multi-step-ahead forecasts. There have been several non-linear statistical tests to quantify the number of inputs for time series such as the Lagrange multiplier (Luukkonen et al. 1988), Likelihood ratio test (Chan and Tong 1986) and bispectrum test (Hinich 1982). These methods are model-dependent, so there is no approach that has any real advantage over another.

The ANN training process is an unconstrained non-linear problem, where the weights are modified to minimise an objective function such as the mean or total square error. Many different optimisation methods exist for ANNs; however, there is no algorithm that ensures a global solution for all problems. The most popular method used for ANN optimisation is the back-propagation method which can be applied using the gradient descent algorithm. Given these limitations of the back-propagation algorithms, many variations of these algorithms have been adopted (Jacobs 1988), (Cottrell et al. 1995). Also, second-order methods have been shown to have faster convergence and ability to find a good local minima, made them appealing and robust training algorithms for the ANN. The limitation of the back-propagation includes slow convergence, inefficiency and sensitivity to the choice of learning rate. Some modifications to the back-propagation algorithm are required, such as the addition of a momentum parameter to overcome issues such as the oscillation problem. Since the learning rate and momentum parameters can take on any value between 0 and 1, the values are determined by experimentation. In Sharda and Patil (1992), nine combinations of three learning rates and three momentum values are examined. Tang and Fishwick (1993) report that these training parameters greatly affect the performance of the ANN. Zaiyong (1991) concluded that a high learning rate should be used for less complex data and low learning rate is more suitable for more complex data. De Groot and Würtz (1991) achieved better major improvements in accuracy and training time for forecasting problems when using second order method. Hung and Denton (1993) used a general purpose non-linear optimiser in the training network. The advantages of this method are that the learning rate and momentum are not used. Instead, stopping criteria, search direction procedure and bounds on variables are configured (Lenard et al. 1995).

The ANN training method is affected by the choice of objective or loss function. The standard approach is to use the sum of squared error or the mean squared error functions. In financial forecasting, other objective functions can be used such as maximising profit or return, or some other utility function. Refenes (1995) show that the ANN predictive performance is affected by the choice of objective function when using the back-propagation algorithm while keeping the learning parameters constant. The use of mean square error is commonly used as the loss function when training ANNs. This assumption is appropriate given the independence of the error term; however, this assumption can be of concern when modelling time series data due to the autocorrelation in the errors. Just like many other forecasting models, the ANN requires training and test data sets. In some cases, a validation set is required to evaluate whether the ANN has reached an optimal solution. The evaluation set can also be used by the training algorithm to avoid over-fitting (Weigend et al. 1990). There is no formal method that dictates how to split the training data set into the training, validation and test sets. The training set must have enough observations to represent the population or underlying process, so by having a smaller set in the training and validation set, this can lead to issues such as selecting the wrong ANN structure, premature termination of the ANN training and inadequate forecast performance. To date, there is no method that dictates how to partition the data set. In the literature, the data set is portioned based on the problem at hand, the most popular methods being 80 versus 20% or 70 versus 30%. Other methods have been used such as (Gorr et al. 1994), where a bootstrap re-sampling method was used to partition the data into ten independent sub-samples. In Lachtermacher and Fuller (1995), all available data is used to train the model then synthetic data series is used to test the ANN. Another interesting issue that arises with time series modelling is the length of the training set, and how to determine the optimal number of observations in the training set such that the optimal model performance is achieved. It is generally assumed that a longer time series is required for training the ANN model to provide adequate performance. In Nam and Schaefer (1995), the ANN forecast performance is tested with different numbers of observations in the training data set. Their study showed improved performance with an increased number of observations in the training set. However, this is solely dependent on the forecasting problem and the complexity and noise in the data. This result contradicts that of (Kang 1992) where ANNs have been reported to perform well when using small data sets compared to linear models.

4.2.2 Neural Networks in Conditional Volatility Forecasting

Neural networks have been extensively applied to time series forecasting problems. However, there has been less work done on modelling volatility with ANNs. This could be attributed to many factors such as the success of the GARCH models in modelling conditional volatility and the complexity of designing the correct neural network for conditional volatility modelling. Conditional volatility forecasting involves modelling the distribution properties of the time series; therefore, the type of neural network model will impact on the forecasting accuracy as it will need to model the distributional properties of the underlying data. The limitation of the neural networks in this regard can be seen in the forecasting literature where the majority of ANN forecasting papers combine ANN with time series models to improve the forecasting performance (Thomaidis and Dounias 2006). To understand this limitation, let us consider that a standard time series can be specified as follows:

$$y_t = f(x_t) + e_t \qquad (4.48)$$

where y_t is the dependent variable, x_t is the independent variable and e_t the residual term which is a Gaussian random error with mean of zero and a constant variance. The ANN would have to perform a non-linear map between the input x_t and the output y_t, which is in fact learning the unknown function $f(x_t)$. This is achieved by increasing the number of the hidden units, so the ANN can approximate any non-linear function to a degree of accuracy as the number of hidden units tends to infinity (Hornik et al. 1989). The training of the ANN is normally achieved by minimising the mean error squared (MES). This approach assumes that the variance of the target conditioned on the input has constant variance (Bishop 1996). That is, the conditional probability distribution function of the target is a single Gaussian of constant variance (Schittenkopf et al. 2000). Bishop (1996) shows that optimising the ANN weights by minimising the sum of error square is achieved by obtaining conditional average of the target data, and a global minimum of the variance of the target data around its conditional average. This form of training has its benefits when modelling the conditional mean of a time series, but this will be problematic when modelling continuous variables such as time varying volatility. Also, the implication of using a loss function such as the Root Mean Square assumes the independence of the error terms, which is not suitable for time series models due to the autoregression in the residuals, (Curry et al. 1997). Therefore, this method is not suitable for modelling volatility in financial time series.

Given the limitation of the standard ANN in modelling continuous random variables, researchers have been able to utilise the benefits of the ANN by combining it with other time series models. In this way, both models can leverage off each other and provide improved forecasting performance. In Donaldson and Kamstra (1996), the output of the ANN is used as an input parameter to another time series model to produce a combined volatility forecast. This method allows for any information that has been missed by the ANN to be captured by the other model. The ANN was chosen to have up to 3 hidden units, where 10 values for each weights of the hidden layer is chosen at random between −1 and 1 and the weights of the output layer is calculated during the optimisation procedure. The set of hidden weights that produces the best ANN performance is used for the forecast exercise. There are several drawbacks of this approach. For instance, the weights in the hidden layer are set in an ad hoc manner; however, the output layer weights are optimised during the training procedure. This method has been proposed by Stinchcombe and White (1992), where guidelines on transforming the data and initialisation of the weights seem to produce good results. However, setting the hidden weights in a random manner without optimisation using the feedback from the output layer can lead to unstable results, especially when applied to financial time series. The output layer weights are the only variables that are optimised in the neural network; therefore, they would need to compensate for the hidden layer weights. Potentially, this method is reduced to a simple regression problem where the output of the ANN is a linear combination of the transformed values of the input layer. This is evident in the results where the forecasting performance of the ANN and OLS are very similar across the different times series. This leads us to conclude that the ANN cannot adequately capture all the characteristics of the time series data, and most of the forecasting ability was driven by the parameters in the GJR model.

Donaldson and Kamstra (1996) made another attempt to combine forecast models using an ANN. In their approach, the output of the ANN is used as an input to the GJR model. This method seems to have produced more stable results than did the previous study. This is not surprising as the ANN output is included in the GRJ variance specification. This means that the forecast of the ANN model as referred to in the paper is actually the output of the modified GRJ model. This method still has the same weakness of the previous study, where the weights of the hidden layer are set randomly and the weights of the output layer are optimised with the combined model. The weakness of this approach can be seen in the results. For instance, three of the four time-series studied use only one hidden unit, whereas three hidden units were used for the NIKKEI time series. Having one hidden unit in the ANN is not sufficient to model the complexities of the financial time series. Hence, from these results we conclude that the fixing of the hidden layer weights has an adverse effect on the ANN's forecasting ability. In the case of the NIKKIE, sufficient information has been missed by the GJR model, which has allowed for the output layer of the ANN to capture this information. This can also been seen from the results, where the ANN is said to behave like EGARCH in the FTSE time series and more like GJR in the S&P 500 time series. By adding extra variables to the conditional variance specification, the model becomes more flexible and adaptable to changes in the underlying data. The key weakness in this research is the way the ANN has been applied to volatility forecasting. This also includes the training method of the neural network where only the linear combination of the output layer was optimised with a conditional variance procedure.

Hu and Tsoukalas (1999) adopt a different combination approach to those of the previous studies. The ANN is utilised to combine the forecast of four other models. The forecast performance of the ANN is compared to the GARCH, EGARCH, IGRACH, simple moving average and another two combine techniques, the OLS and a simple average method. The models are trained on 11 currencies covering the first 15 years of the European Monetary System. The ANN hidden layer is selected to have four hidden units, and four skip connections connecting each input node to the single output node. The results of these experiments show the superiority of the EGARCH model over all models including the ANN. The ANN displayed poor performance compared to all other models which is attributed to the structure of the ANN used to combine the forecasts, where four hidden units were used without verifying whether this was the optimal structure. The choice of a skip connection in the ANN was not supported by appropriate validation. The input to the ANN consisted of volatility forecasts of different models. This ANN would not be able to predict volatility solely by using the volatility forecast. This issue could have been overcome by adding information about the return time series to the ANN.

Dunis and Huang (2002) applied a neural network recurrent (NNR) and regression neural network (RNN) to forecast foreign currency volatility. They benchmarked both models against the GARCH(1,1), implied volatility model, two types of combination models, a simple average of the four forecasting models, and the GR model proposed by Granger and Ramanathan (1984). The data used in the study is the USD/JPY and GBP/USD exchange rates. It seems that both return series have zero unconditional mean, and the USD/JPY return series seems to be homoscedastic, whereas the GBP/USD seem to be heteroscedastic. The inputs to the neural networks are the lagged actual 21 realised volatility, lagged 21-day implied volatility, lagged absolute logarithmic return of the past 21 returns and lagged logarithmic returns of the past 21 returns of the gold or oil prices. The target of the neural networks is the realised 21st day volatility. The results show that the RNN dominates the NNR, highlighting the importance of the feedback effect of the lagged variables as inputs. The NNR also dominated the GARCH in terms of forecasting performance. However, a comparison of the n-step-ahead forecast of the GARCH and the neural networks would be misleading. If the volatility persistence is weak in the data series, then the GARCH model could revert to the unconditional volatility in the long run. It would have been a better exercise to compare the ANN with a more sophisticated version of the GARCH model such as the EGARCH or GJR models. This becomes evident when applying ANN in trading strategies where the GARCH model produces higher cumulative profit compared to the RNN and NNR models. However, for the USD/JPY, the NNR and RNN models produced better results than the GARCH. This is expected as their return series is not heteroscedastic. In this case, it would have been ideal to compare models with an ARIMA model.

In Hamid and Iqbal (2004), a primer on neural networks in volatility forecasting was provided where issues involved with volatility forecasting are explained with references to the literature. They also test the forecasting capability of the ANN to forecast the volatility of the S&P 500 Index future prices, and benchmark it against American option implied volatility model. The ANN was trained using thirteen input variables to learn the realised volatility. The number of hidden units was set to twenty-six; no test was done to validate this choice. The other ANN parameters such as learning rate and momentum were set at the default value. The ANN was trained on 500 days of data, which is a common oversight for the length of the in-sample or the training data set length to be chosen in an ad hoc manner. The results show the ANN to have better forecasting performance than the implied volatility model. This test is not sufficient to conclude the superiority of the ANN, since the implied volatility is forward looking which implies how the market is anticipating the volatility to behave. The ANN was trained on the historical data or realised volatility that is backward looking where the volatility is forecast based on the past behaviours. So in essence, it would have been more informative to compare the forecasting ability of the ANN to time series volatility models such as the GARCH models.

Thomaidis and Dounias (2006) introduce a non-linear method in a mean neural network with a GARCH specification for the variance (i.e. NN-GARCH). Based on the maximum likelihood theory, they devise statistical inference procedures for the NN-GARCH model, which allows for the testing of the structure of the mean and variance data generating conditional probability density. The idea is to have a flexible mean specification and a GARCH specification for the variance to cater for a good fit for the return time series. The inputs used for the ANN are the last 10 lags of the DAX returns. Two or three hidden units were selected as part of the ANN modelling building procedure. The forecasting results are inconclusive. This method might give a good specification from the conditional mean; however, it is still dependent on the variance to follow a GARCH specification. Nevertheless, the proposed method seems to have some advantages and can be adopted for time series modelling building.

Xiao et al. (2008) conducted a similar experiment to that of (Hu and Tsoukalas 1999), whereby the ANN was used to combine the forecast of GARCH, EGARCH, TGARCH and moving average volatility models. The ANN model was compared to these models including others such as EWMA, exponential smoothing and moving average models. The ANN was also compared with two other combining models, the standard average and OLS models. The ANN was constructed with four inputs which are the outputs of GARCH, EGARCH, GARCH and moving average models, one hidden layer with four hidden units, and output unit that is the combined forecast. The models were trained on the Shanghai Stock Exchange composite closing index. The training data set consisted of 382 observations of weekly volatility; the testing was done on 100 weekly observations. The results show superior performance of the ANN combining technique compared to the other volatility models. Also, it is interesting to see the OLS method achieve very similar results to those of the ANN. This shows that the power of neural networks has not been explored to its full potential, where the ANN is acting as a regression function which is caused by the ANN architecture and choice of inputs.

Ormoneit and Neuneier (1996) applied conditional density estimation (CDE) where the parameters of the probability density function where generated by ANN. In this study, the neural network is trained using DAX index intra-day returns. The aim was to predict DAX volatility 60 min ahead, based on the present value and 4 lagged values the 20, 50, 90 and 150 min. The data set consisted of 2000 observation for training and another 1585 observations for testing. The results of the CDE were compared to a standard MLP and a constant predictor. The CDE was shown to have superior volatility forecasting performance over the ANN, however, it would have been deal if the performance of the CDN is benchmarked against other time series models such as the GARCH or EGARCH models. This would have allowed for a complete evaluation of the results and would have highlighted the strengths and weaknesses of the proposed model.

The majority of neural network research in financial forecasting applies a hybrid approach. This involves combining the neural networks with other time series to improve on the forecast. This is surprising as neural networks have the ability to map the inputs to a target value more accurately than financial time series models understudy. Also, the neural networks do not apply the same assumptions as the statistical models on the data set. The reasoning for this approach is the inability of the feedforward (MLP) neural networks to capture key features in the underlying data such as heteroscedasticity in the data. We demonstrate in this book the capability of neural networks to achieve accurate forecasting when the correct model is chosen.

4.2.3   Volatility Forecasting with MDN

In Bishop (1996), the MDN were shown to be capable of modelling the complete conditional density function of the target variable, which is an essential property for modelling time series data. In addition, the MDN variance specification caters for changes in the variance relative to the input variables. This is a major advantage over ANN where the global variance specification is a constant over the entire population. Designing the correct model for forecasting volatility is complicated by the stylised facts that exist in the return series. The MDN can potentially address this issue due to its flexibility and its ability to capture key behaviours in the underlying data. MDNs have not received the required attention in research especially in the forecasting domain. To date there is a handful of papers which utilise MDN in volatility forecasting problems. Below is summary of these papers.

In Nabney and Cheng (1997) the volatility forecasting performance of ARIMA model was compared with different types of ANN namely MLP, MDN and RBF. The data set included five currencies with 3505 daily prices. The five previous prices were chosen as the inputs for the neural networks and for the price of the next time step, the target for the neural networks was the conditional variance. The ANNs are trained on the first 2505 observations; the next 500 are used for the validation and last 500 for out-of-sample testing. Their results demonstrate the superiority of the MD relative to the other models. However, this research does not compare the performance of the neural networks to other time series models.

Schittenkopf et al. (2000) extend the MDN so that it has a recurrent structure (rMDN). This is achieved by feeding the past variance as an input to the MDN model. Now the past volatility observations can influence the future volatility forecast. The rMDN was selected to have three hidden units and two Gaussians. The rMDN performance is compared to the GARCH and GARCH-t models. The models were trained on the daily closing prices for the DAX index. The training data set consisted of 1700 observations. The data set was split into eight overlapping segments that contained 1000 observations. The models were optimised on 900 observations and the next 100 days was used to evaluate the forecasts. The forecast evaluation criteria are based on the loss function (log likelihood), normalised mean average error (NMAE) and the hit rate (HR). The in-sample results show the rMDN to have superior performance over the GARCH models. This is a great improvement on the results of their earlier study, and highlights the importance of selecting the right neural network model to address the problem at hand. In terms of the out-of-sample forecast performance, the GARCH-t on average is superior to all the other models. The rMDN was able to beat the competing models on several occasions. However, the performance seems to deteriorate with large error in the data set. They concluded that the performance of the model depends on the error measure and the section of the time series studied. This highlights the importance of selecting the correct training set for optimising the model which influences the model forecasting performance. The poor performance of the rMDN could be attributed to the model design; for instance, the three hidden units and two Gaussians might not be sufficient to model a time series with large volatility periods. Nevertheless, the rMDN was shown to have a high correlation with the implied volatility when compared with the GARCH-t using the Pearson's r test. This result warrants further research into MDN.

Miazhynskaia et al. (2003) extend the study of Schittenkopf et al. (2000) to include three time series, Dow Jones Industrial Average (DJIA), FTSE 100 and the NIKKEI 225. The data consisted of 13 years of daily closing prices. The data set was split into overlapping sequences of 700 observations. The first 500 returns were used for training the model, the next 100 was used for validation and the other 100 for out-of–sample testing. The models studied in this paper are GARCH(1,1), RMDN(1), RMDN(2)-t, GARCH(1,1)-t, LRMDN(2), and RMDN(2). The RMDN (2)-t is a MDN with a Student-t instead of a Gaussian distribution. The LRMDN (2) is also an MDN where the hidden units are linear activation functions. Three hidden units were chosen for all the neural networks. The models were ranked according to the log likelihood, NAME, NMSE, HR and weighted hit rate (WHR). The error measures of the models vary with the data segment and the difference between the error measures is actually small. So a Wilcoxon test is applied to detect the correlations between the errors. The out-of-sample results for the DJIA showed the GARCH(1,1)-t to be superior to other models according to the loss function measure. The RMDN(1)-t seem to do better according to the volatility measures. For the FTSE 100, the GARCH(1,1)-t and LRMDN(2) outperform the other models with respect to the loss function. The other error measures favour the t-distribution and the mixture of Gaussians over the other Gaussian models. For the NIKKEI 255 series, results obtained are similar to those for the FTSE 100, where the t-distribution models are favoured over the other models. These results demonstrate the superiority of the non-Gaussian models in terms of the forecasting performance. However, the non-Gaussian models perform differently across different time series. For instance, for the DJIA, the t-distribution gave superior performance while the mixture models were among the best for FTSE 100. Even though the mixture and t-distribution models are able to capture fat tails in the return series, only the MDN are able to cater for time varying skewness and kurtosis. These results highlight the importance of selecting the right forecasting model for the time series understudy. It would have been interesting to see if any performance gain would have been achieved by varying the number of the hidden units for the MDN models.

MDNs are suitable for volatility forecasting as they are able to capture the key stylised facts in the returns series such as heteroscedasticity. However, there have been a handful of papers that study the forecasting capabilities of the MDNs. The results reported in these papers are inconsistent and they show different behaviours of the models for the same and different time series. This behaviour is caused by the many important factors that should be considered as part of the model building exercise. For instance, the neural network training data is chosen in an ad hoc matter which is the common approach in time series models. The other interesting factor is that the number of hidden units in the neural networks is not verified or optimised accordingly. No attempts have been made to explore the behaviour of the same model across different time series. In this book, we will elaborate on these key factors and how they impact on the accuracy of the models. This will help us to explain this inconsistency in the reported results.

4.2.4   Application of Neural Networks Option Pricing

Many option pricing models have been derived to capture the relation between the option price and the variables determining it. To model the dynamic relationship between these variables and the option price, a model is required that is agile and flexible enough to cater for the finer details in the underlying data. Neural networks are well suited for this task due to their ability to approximate any measurable function with an arbitrary degree of accuracy. Another advantage is that neural networks do not have any prior assumption about the data. Therefore, the model is data-driven so it is not forced to conform to a specific configuration.

One of the first attempts to apply neural networks to price option was made by Hutchinson et al. (1994). In this paper, they propose a nonparametric method for estimating the derivative pricing formula by using neural networks. The neural network is explored since it could yield more accurate results and might be computationally more efficient especially when the pricing dynamics are unknown or when there are no closed form solutions for the pricing formula. Four pricing methods are used in this study: ordinary least squares, radial basis functions networks, multi-layer preceptron networks and a projection pursuit method. The inputs for the neural network are the stock price, strike price and time to maturity. The volatility and risk free rate are assumed to be constant for the life of the option; therefore, they are not included in the training data set. The inputs are simplified further by using the homogeneity hint (Merton 1973), where it is assumed that the distribution of the asset returns is independent of the level of the stock price. So the option pricing formula is homogenous of degree one in the stock price and strike. This potentially can reduce the inputs to stock price/strike and time to maturity, but this assumption is relaxed and the stock price is included as an input, and the ration option price/strike was selected as the target. All neural networks had four hidden units which were selected using simulation results, and the number of hidden units was based on the smallest out-of-sample error. The evaluations of all models were based on the hedging performance according to the Black-Scholes formula using the tracking error. The evolution was first conducted using Monte Carlo simulation to simulate the asset paths, then calculating the tracking error accordingly. The tracking error varied with respect to the maturity and strike. On average, the RBF network had superior performance according to the Black-Scholes for at-the-money options across all maturities. As the maturity increases, on average the performance of the RBF network improves for deep out-of-the-money options. The average performance of the MLP and PPR networks is on par with that of the RBF network with respect to all maturities at-the money and longer maturities for deep out-of-the-money options. The linear models in rare cases outperformed the Black-Scholes in daily hedging scenarios. It is also noted that the neural networks' pricing performance should be monitored for extrapolation, since they are trained on a specific region of the input space. Therefore, it should not be surprising that the neural network does not perform as well for points outside the input space. This compromise in neural network performance is due to the inadequacy of the neural network structure such as insufficient number of hidden units or the inadequate size of the in-sample training data set. The neural networks showed promising results for the simulated data, so the experiment is extended to S&P 500 futures options. The data set of S&P 500 futures option comprised the daily closing prices for the January 1987 to December 1991 period. The data set was split into 10 non-overlapping subsets, and each neural network was trained on each data set and tested on the subsequent set. The optimal neural network structure is chosen to match the optimal structure used in the simulation experiment. In general, the neural networks exhibit smaller hedging error than does the Black-Scholes, but it is difficult to infer which neural network is superior. For short maturity options, the Black-Scholes had lower hedging errors across all moneyness; however, the neural networks were superior for medium- and long-term maturities.

The pricing accuracy of the neural network is studied in Lajbcygier et al. (1995) and compared with the Black-Scholes and Barone-Adesi/Whaley models. The study was conducted on the SPI futures and SPI options on futures data obtained from the Sydney Futures exchange. The data ranges from January 1993 to December 1994. Eighty percent of the data was used as in-sample; the rest was used for out-of-sample testing which was selected randomly. Several neural networks were tested on the training data sets using different configurations to compare with the results of (Hutchinson et al. 1994); mainly, the assumption of constant volatility and interest rate is tested. Therefore, two sets of inputs were used which consisted of moneyness (stock/strike) and time to maturity. The second set included interest rate and volatility. For the first input set, two neural networks were used with 4 and 10 hidden units. For the second input set, three neural networks were applied using 4, 10 and 20 hidden units respectively. A logistic activation function is used in the hidden and output layer. The performance of the models is evaluated based on three criteria: R^2, NRMSE and MAPE. The results show superior performance when interest rate and volatility are included. The neural network results overall are comparable to those of the other models, although it is marginally out-performed over the entire set. It is interesting to note that the Black-Scholes is superior in the R^2 measure, whereas the NRMS favours the Whaley model. These results show the danger of relying on one measure which could lead to a wrong conclusion when benchmarking different models. The experiment is repeated for a reduced input space that includes near-the-money options that have a short time to expiry. The neural network displays superior performance to the other models. The main weakness of the study is the random selection of observations for the out-of-sample data set. This potentially has a negative impact on the forecast performance due to discontinuities in the training data set. This effect is further analysed and discussed in Yao et al. (2000).

In Garcia and Gençay (2000), performance enhancement in derivative pricing is investigated using the homogeneity hint. Instead of using a neural network to map the strike to stock ratio and time to expiry to a derivative price, the pricing function is split into two parts whereby one is controlled by time to maturity and the other by the strike to stock price ratio. This separation is consistent with the BSOPM which retains the homogeneity property of the option pricing formula. The experiment consists of training two feedforward neural networks, one with the homogeneity hint and the other without the hint. The number of hidden units by experimentation and the optimal number were selected based on the lowset pricing error. It was interesting to note that the number of hidden units for the neural network without the hint is higher compared to the neural network with the hint. The initial simulation results were encouraging, the neural network with the homogeneity hint displayed superior pricing performance; however, the neural network without the hint produced a slightly lower hedging error. The experiment was repeated for a real data set using the daily S&P 500 index European options (obtained from CBOE). The data sample was from January-1987 to October-1994. Each year was split into three sets: the first set consisted of the first half of the year for training purposes, the second set was a validation set which consisted of the following quarter and the third set was the final quarter of the year which was used for out-of-sample testing. Based on the pricing error criteria used in the mean squared prediction error (MSPE), the neural network with hint consistently produced fewer errors than did the neural network without the hint and Black-Scholes. Based on the simulation experiment, the model that produces the least number of pricing errors does not necessarily produce fewer hedging errors. Therefore, for the hedging scenario, the neural network hidden units are selected based on the lowest hedging criteria rather than the MSPE. The results indicate similar hedging performance for both neural networks. However, the neural network with hint seems to be less sensitive to the initial condition and on average produces more stable hedging results. Also, the hedging performance of both neural networks is superior to that of the Black-Scholes.

Lajbcygier and Connor (1997) introduced a hybrid neural network which predicts the difference between the modified Black-Scholes and the observed option prices. The residual predictions are applied to supplement the optimal conventional option pricing model out-of-sample. The neural network input consisted of future price to strike ratio, time to expiry and volatility. The target data set consisted of the residual between the modified Black-Scholes model price the option price. The data used in this research is the intra-day option from Sydney Future Exchange for the period of January 1993 to December 1993. The first half of the data is used for training and the rest was used for out-of-sample testing. The neural network selected had fifteen hidden units and it was optimised using back-propagation with 20% validation set used for network selection. Confidence intervals were derived using bootstrap methods which are used in trading strategies that allow only those trades outside the estimated range of spurious model fits to be executed. The hybrid model seems to improve option pricing; however, there is still pricing bias which was addressed by using bootstrap methods.

Hanke (1999) conducted an empirical study on neural networks methods for pricing options. The neural networks were designed to have moneyness, volatility and time to expiry as inputs; option price to strike ratio is used as the target. Another neural network was constructed which utilises the hybrid approach used in Lajbcygier et al. (1997). This method replaces the target value of option price divided by strike with difference of the option market value with the BSOPM value divided by the strike. The data used in this study are DAX call options for the period April 1994 to Feb 1995. The data was divided into 120 days for training, 20 days for validation and 50 days for testing. To select the optimal structure, first the neural network is selected to have two hidden units then the number of hidden units is increased by two until the number of hidden units reached ten hidden units. Results show the standard method is inferior to the BSOPM, whereas the hybrid approach shows superior performance to that of the BSOPM. The results are listed as total averages which are not broken down by the option moneyness. So it is not clear how the hybrid approach fares across moneyness and maturities in terms of the BSOPM.

In Schittenkopf and Dorffner (2001), MDN is used to extract the risk-neutral density from option prices. The concept behind this approach is to utilise the MDN capabilities to model the shape of the risk-neutral density as a function of the maturity of the option. The FTSE 100 European options data for the period of 4th/January/1993 to 22nd/October/1997 was used. The data was first checked for boundary conditions and recording errors. Also, contracts with maturity less than two weeks, with price below 5 points, moneyness outside the rage of [−0.1, 0.1], and annualised volatility above 50% or below 5% are removed. The final data set contained 65549 option contracts. The input consisted of the time to expiry of the option and the output was the actual option price. The MDN was trained using a sliding window of 10 days, and minimising the mean squared pricing error using the conjugate gradient descent method. The MDN was benchmarked against the BSOPM and the Black-Scholes adjusted for skewness and kurtosis. The pricing performance was evaluated using the pricing error, and the absolute pricing error. The results show the overall pricing performance of the Black-Scholes is comparable to the MDN, for at-the-money and maturity in the range of short to medium horizons. However, the MDN shows superior performance for long maturities across all moneyness categories. This is also valid for out-of-the-money and in-the-money options, which is due to the distributional generalisation of the MDN. In terms of hedging performance, the MDN was worse-off compared to the other models. This could be due to having the option maturity as the only input, which potentially can prevent the MDN from modelling key dependencies between the maturity and moneyness with the option price. Also, the use of puts and calls in the same training set without an indicator in the input set could skew the results.

An interesting method adopted by Hanke (1997) utilises the neural networks to approximate option prices and the option Greeks. This is particularly important in the case where no analytical solution exists. The idea behind this approach is to train the neural network on a carefully selected artificial data set that covers a wider range compared to the real market data. The data is generated by using a range for each variable; for instance, the volatilities for certain markets fluctuate between 0.1 and 0.4. This will limit the training data set to an acceptable and valid range. The output or target pattern is calculated analytically using GOPM. The neural network architecture consisted of seven input variables, 50 hidden units and one output unit. The layers were fully interconnected, and the logistic functions were used as the activation functions for hidden and output units. The optimisation of the neural network was achieved by minimising the sum of squared difference between network output and target. The out-of-sample performance of the neural network seems to have higher deviation for in-the-money options, since the majority of traded options fall into this category which is an undesirable bias. To overcome this issue, a hybrid approach is used similar to the approach of (Lajbcygier et al. 1997), where new patterns are used which consist of the difference between the simulated prices and the analytically computable price close to them as the new output pattern. For example, the difference between simulated price and the standard BSOPM price is used to train the neural network. The aim is to allow the neural network to learn the relationship between the input space and the difference between GOPM and the Black-Scholes model. The out-of-sample results show a higher accuracy and reduce the undesirable bias. However, the difference between the delta formulas and the Black-Scholes deltas lead to a higher RMES. This could be due to the lack of representation for the delta in some particular ranges which causes the neural network to misrepresent those categories. This issue was not resolved and is left for further research. This method is an attempt to overcome the simulation and computation issues of models that do not have a closed form solution such as the GOPM. However, it is not clear if this method holds under extreme market conditions, and whether it can produce consistent results across different time series for different periods.

In Yao et al. (2000), experiments are carried out using neural networks with the aim of overcoming the drawbacks of the BSOPM. The input data set included the stock price, strike price and time to maturity. The volatility and interest rate are omitted from the data set since they are assumed to remain constant over the life of the option. The neural network is trained using the option price as the target. The data set used is the Nikkei 255 stock index options traded on the Singapore International monetary exchange. The data sample ranges from 4 January 1995 to 29 December 1995. The data set is split using 70, 20 and 10% as the training, validation and testing data sets accordingly. The back-propagation neural network consisted of one hidden layer with sigmoid hyperbolic tangent for the activation functions. The neural network was optimised using the normalised mean error squared as the loss function. The number of hidden units was selected based on the number of weights that produces the lowest NMES. The experiments included the use of different neural networks with different in-sample data sets. The training set was applied in many forms including: sorting the data according to trading date, sorting the set according to the time to maturity, randomly selecting data and partitioning the data according to the moneyness. The initial results favour the data to be sorted according to the trading date as this allows the neural network to capture the volatility dynamics even when it is not included as an input. Also, the results show pricing improvements when the data is partitioned by moneyness and a different neural network is used for each data set. The neural network pricing performance is superior to that of the Black-Scholes for in-the-money and out-of-the-money; however, it does worse for at-the-money options.

Amilon (2003) applies neural networks to derive the bid and ask prices rather than assuming the midpoint or the last traded price. The neural networks' pricing and hedging performances are compared to that of the BSOPM using historical volatility and implied volatility estimates. The data used in this study are the daily closing quotes of the Swedish OMX index and daily closing bid and ask prices of European OMX index call options ranging from June 1997 to March 1998 and June 1998 to March 1999. The reason for this split is to avoid adjustments for the index dividends. The MLP networks with a different number of hidden units are trained on the first four months, and then validated on the following two months. The neural network with the lowest errors is selected and evaluated on the following month. The process is then repeated by using the first five months for the training data set. Twenty different neural networks were trained using 10, 12, 14 hidden units for each for the eight training sets. The results show the neural network pricing that is derived from the option data has superior performance compared to the other models. In terms of hedging, the neural network trained on the asset price time series is superior to the other models. The results do not report the best performing neural network for each training set. In this research, the neural network that best fits a particular data set is used. Hence, the performance of the neural network is very much dependent on the neural network applied for that section of the time series. However, the ideal model should be able to categorise data across different time periods. One cannot conclude from this research that the neural network is superior to the BSOPM since different neural network models are used for different time periods.

Bennell and Sutcliffe (2004) compare the pricing performance of the BSOPM relative to neural networks for the FTSE 100 European call options traded on LIFE for the period ranging from January 1998 to March 1999. The data set is split into two sets according to the options moneyness (in-the-money and out-of-the-money); for each set, one third was used for validation. Different neural networks were trained on each data set which involved searching for the right combination of inputs, hidden units and outputs. Each of the neural networks was trained 100 times and the best performing network was selected. This is to reduce the dependence on the initial network weights values. The models were benchmarked using correlation (R^2), mean deviation, mean absolute deviation, mean proportionate deviation and mean squared deviation. The results show that the homogeneity hint greatly improves the pricing performance of the neural networks compared to using the stock price and strike separately in the input space. Also, training different neural networks on each moneyness category greatly improves the performance of the neural network. For out-of-the-money options, the neural network is superior to the BSOPM. For in-the-money options, the BSOPM model outperforms the neural network where the neural network severely misprices long time to maturity in-the-money options.

Meissner and Kawano (2001) constructs different neural networks to price options by using a consisted volatility input. The volatility is derived through the GARCH model which is used an input to the neural network. Four types of neural networks are compared to the BSOPM based on the pricing performance. The models were tested on ten technology stocks. The data set was partitioned into training, validation and testing portioned at 2:1:1 respectively. The four neural network models used are MLP, RBF, Generalised regression networks (GRNN) and linear network. The inputs consisted of moneyness, time to maturity, risk free rate and the volatility derived from the GARCH model. The number of hidden units and layers were optimised for each data set. The results show that the MLP is superior to all models on all stocks; also, all neural network models outperform the Black-Scholes on all measurement criteria for each stock. However, the GRNN was superior when all stocks data were combined. The analysis displayed only the average performance of the models; no further analysis was done on option pricing according to maturity and moneyness.

Samur and Temur (2009) apply neural networks for option pricing on European and American options. The training data consisted of 134 of S&P 100 European and American Index put and call options. The time period seems fragmented where different days were used to accumulate the data set. Three MLP neural networks were used, the first neural networks inputs consisted of option type (American \European), strike price, spot price, maturity and interest rate. The second neural network included the variance in the input were the variance was calculated as the monthly volatility, and the third neural network variance was the annual volatility. The volatility inputs did not to improve the pricing accuracy. This is surprising since a lot of other research has established the importance of the volatility parameter. The only issue arising from this is that the in-sample data set is not continuous in time, which seems problematic especially when using put and call for European and American options in the same data set. The other issue is that the use of 134 data points to train the neural network is insufficient given the large number of variables in the neural network.

It is interesting to see that the majority of the research evaluates the neural network option pricing accuracy in relation to the BSOPM. However, the BSOPM's weaknesses and drawbacks are well documented in the literature. Few, if any, papers have attempted to evaluate the pricing capabilities of more advanced option pricing models. The reason for this is the pricing capabilities of the neural networks are inferior to the other advanced models. In this book, we elaborate on this behaviour and provide a new method that overcomes this drawback.

Chapter 5
Important Problems in Financial Forecasting

In this chapter, the problems addressed in this book are defined in a clear and concise manner. We start by defining the terms and concepts used and how they are used throughout the book. Then we discuss the methodology applied to solving the problem as stated in the problem definition.

5.1 Terms and Concepts Used

This research covers four main areas of market risk which are: time series forecasting, option pricing, hedging and value-at-risk. The focus of this book is on the application of neural networks in market risk. That is how neural networks can be applied to market risk to overcome the limitation of existing financial models. There are many terms and definitions that have different interpretations, so we provide a definition and clarify the usage of these terms in the context of this book.

5.1.1 Financial Time Series

In this research, time series forecasting is referred to as 'financial time series'. The financial time series applied in this research include stock prices, returns, option prices etc. Forecasting financial time series is an integral part of market risk, where investors need to understand the future movements in order to make informed decisions. Forecasting financial time series is complicated by the element of uncertainty in the financial theory and empirical time series. Therefore, statistical theory is essential in developing financial models and plays an integral role in financial time series forecasting.

© Springer International Publishing AG 2017
F. Mostafa et al., *Computational Intelligence Applications to Option Pricing, Volatility Forecasting and Value at Risk*, Studies in Computational Intelligence 697, DOI 10.1007/978-3-319-51668-4_5

5.1.1.1   Underlying Asset

Underlying asset specifically refers to the stock or index traded on a stock exchange.

5.1.1.2   Returns

The statistical properties of the return series are more aligned to measure potential loss and profits of a portfolio which makes them more attractive than asset prices in financial risk management. Also, the returns series is a complete and scale-free summary of the investor's investment opportunities.

'Returns' in this book are defined as follows:

Let P_t be the price of the underlying asset at time t, assuming asset pays no dividends the return is can be calculated as the continuous compounded return r_t,

$$r_t = \ln\left(\frac{P_t}{P_{t-1}}\right) \tag{5.1}$$

Modelling the return series is simplified using Wold's decomposition. This return series consists of deterministic and non-deterministic components and is represented as follows:

$$r_t = \mu_t + \sum_{j=0}^{\infty} b_j \varepsilon_{t-j} \tag{5.2}$$
$$\text{where,} \, b_0 = 1_t$$

where μ_t is the deterministic part of the return series and is normally modelled using ARIMA models. The non-deterministic component ε_t is the uncorrelated sequence residuals which is known as the innovation of the process r_t, b is the infinite vector of moving average weights or coefficients, which is absolutely summable accordingly, i.e. $\sum_{j=1}^{\infty} |b_j| < \infty$.

5.1.1.3   Return Stylised Facts

The return series of an individual stock display patterns over time. These patterns and behaviours are also common across different financial time series. These common patterns and behaviours are referred to as 'return stylised facts'.

5.1.1.4 Conditional and Unconditional Mean

The unconditional mean of the return series r_t is the weighted average for the entire return series.

$$E[r_t] = C \tag{5.3}$$

where C is a constant.

The unconditional mean can is defined as

$$E[r_t|\phi_{t-1}] = \sum_{i=1}^{\infty} b_i \varepsilon_{t-i} \tag{5.4}$$

where ϕ_{t-1} is the information set such that $\phi_{t-1} = \{\varepsilon_{t-1}, \varepsilon_{t-2}, \ldots\}$.

5.1.1.5 De-meaned Return Series

A de-meaned return series is referred to as a 'return series of mean zero' where the condition mean has been subtracted from the series.

5.1.1.6 Conditional and Unconditional Variance

The conditional variance of a return series r_t is calculated as follows:

$$Var[r_t] = \sigma_\varepsilon^2 \sum_{j=0}^{\infty} b_i^2 \tag{5.5}$$

The conditional variance of return series r_t is given by:

$$E[(r_t - E[r_t|\phi_{t-1}])^2 | \phi_{t-1}] = \sigma_\varepsilon^2 \tag{5.6}$$

5.1.1.7 Volatility

The volatility of the return series is not directly observable in the market; therefore, the volatility measures have to be approximated. To understand how the returns series will behave in future, it is essential to devise a model that can filter the volatility measure using the current market information. The common practice is to convert the asset price series to return values which are more suitable for this task. Then statistical models can be formulated to forecast the returns series which involve modelling the conditional mean and variance based on the historical return series. The models will be able to forecast future volatility measures, where the forecast is determined by the information available at the time and is very much dependent on the historical data. In a general sense, volatility is calculated as the variance of the return series which measures the dispersion of the density function around the mean. The variance measure on its own is a distribution-free parameter; only when it is associated with a standard distribution can the probability density and cumulative density be analytically derived (Poon 2005). Calculating the volatility of return series using the variance Eq. 3.5 would result in a constant volatility measure which is known as the 'unconditional volatility'. This measure is commonly used in finance models due to its simplicity. However, it not suitable for volatility forecasting as it is not a true representation of the observed volatility in the return series. The volatility approximation process is further complicated by the return stylised facts. This means that even if we attempt to use the conditional variance as a proxy for the observed volatility, this will still not cater for the some of the key stylised facts observed in the return. Therefore, we need to define volatility to maintain consistency in our approach.

Definition
Volatility is the conditional variance of the de-meaned return series such that the evolution of the volatility through time is governed by a defined stochastic process and is associated with a well-defined probability distribution.

5.1.1.8 Conditional Mean Forecasting

Conditional mean forecasting is the process of determining the future expectation of a return series. The forecasted return values are dependent on the information present at time t. See Appendix A, Sect. A.4.

5.1.1.9 Conditional Volatility Forecasting

In this book, conditional volatility forecasting is referred to as 'volatility forecasting' unless otherwise specified. Due to the wide variations of variance specification for volatility forecasting, the volatility forecasting must be defined in terms of the volatility model used throughout this book, such as GARCH.

5.1.2 Options

An option is a contract between two parties, the buyer and seller. The buyer purchases from the seller the right but not the obligation to buy or sell the underlying asset at a fixed price in a given time frame. The buyer has to pay the seller a fee (a premium) for the purchase of the option.

5.1.2.1 Strike Price

Strike price of an option is the fixed price that the underlying asset would be delivered when the option is exercised.

5.1.2.2 Call Option

A call option gives the buyer the right but not the obligation to buy the underlying asset at the strike price in a set time frame. The pay-off function of the call is given by

$$Call\,Payoff = \max(K - S, 0)$$

Where K is the strike price and S is the price of the underlying asset.

5.1.2.3 Put Option

A put option gives the buyer the right but not the obligation to sell the underlying asset at the strike price in a set time frame.

$$Put\,Payoff = \max(0, K - S)$$

Where K is the strike price and S is the price of the underlying asset.

5.1.2.4 European Options

European style option can be exercised only at the expiry date of the option.

5.1.2.5 Moneyness

The options' moneyness is determined by the value of the asset relative to the strike price of the option. So, if the an option is exercised and an immediate positive cash flow is realised, that is the stock price is greater than the strike price (for a call), then the option is said to be *in-the-money*. If the strike price of the option is equivalent to the stock price then zero cash flow will be realised; this is referred to as *at-the-money* option. If the strike price is greater than the stock price then if the option is exercised, a negative cash flow will be realised. This option is referred to as *out-of-the-money* option.

5.1.2.6   Risk Free Rate

The risk free rate is the rate of return when having zero risk. Typically, the government treasury bills are used as a proxy for the risk free rate.

5.1.2.7   Option Pricing Model

The option pricing model produces an analytical option price which is referred to as the 'fair option price'. This analytical option value is then compared with the market price of the option to determine if the option is overpriced or under-priced. The option pricing model accepts different inputs and in most instances makes some assumptions to allow for the derivation of the model. These assumptions typically include the following:

- Instruments can be bought and sold freely in the market
- Information is readily available to all participants; i.e. the market is efficient
- There exists a risk-free bond in the market
- There is no arbitrage
- The market is complete—agents can exchange goods directly or indirectly with each other

5.1.2.8   Implied Volatility

The implied volatility of an option is the volatility value of the underlying asset that is implied by the option price based on an option pricing model. That is, the volatility value when plugged into an option pricing model would yield an analytical value identical to the market option price. In this research, the implied volatility parameter is defined in terms of the BSOPM, where the implied volatility, $\sigma_t^{BS}(K,T)$ is a function of K (Strike) and T (time to maturity).

$$C_{BS}\left(S_t, K, \tau, \sigma_t^{BS}(K,T)\right) = C_t^*(K,T)$$
$$\text{where } \sigma_t^{BS}(K,T) > 0 \tag{5.7}$$

5.1.2.9   Hedging

Hedging refers to the use of options to offset or mitigate risk that is attributed to fluctuations of prices. Typically, this is achieved by calculating the sensitivities of an option according to a certain parameter using an option pricing model and purchasing the equivalent amount in the underlying asset or placing the equivalent amount in the bank.

5.1.2.10 Delta Hedging

Delta hedging specifically refers to hedging the risk of the movements in the underlying asset. This is accomplished by calculating the partial derivative of the option price relative to the underlying asset:

$$Delta = \frac{\partial Option}{\partial Stock}$$

5.1.2.11 Value at Risk (VaR)

In simple terms, the VaR measure makes the following statement: for the N days we are $X\%$ certain that we would not lose more than M dollars. This summarises the total portfolio risk exposure to a single number that can be easily understood by risk managers. The calculation of the VaR measure is based on the profit and loss distribution assumption. This is typically assumed to be a normal distribution for ease of calculation.

Definition
Assuming normality in the return series with confidence interval α, return volatility σ, initial portfolio value W0 and time horizon N days,

$$VaR_\alpha = \alpha \sigma W_0 \sqrt{N} \tag{5.8}$$

5.2 Problem Definition

The problem that will be addressed in this book is the current lack of a neural networks forecasting model that is suitable for:

- Volatility forecasting
- Option pricing
- Delta hedging

In this book, we propose to develop methods that effectively provide forecasting of these three market elements.

5.2.1 Volatility Forecasting

We will carry out forecasting of conditional volatility of different return series. Since the focus is on the conditional volatility, the return series used will be de-meaned which will simplify the forecasting problem. The condition volatility forecasting problem is achieved through time series modelling techniques.

$$r_{t+1} = \mu_t + \varepsilon_t$$
$$\text{where} \quad \mu_t = E(r_t|\Theta_{t-1}) \text{ and } \sigma_t^2 = Var(\varepsilon_t|\Theta_{t-1}) \tag{5.9}$$

Θ_{t-1} is the information available at $t-1$. In this book, the conditional volatility forecast can be specified by an exact function or stochastic equation that dictates and governs the evolution of the volatility process σ_t^2.

5.2.2   Option Pricing

In this book, we will price options using different option pricing models. The accuracy of the models is very much dependent on the model assumptions and the quality of the data used. The option pricing formula produces a fair option pricing according to the following,

$$O_t = P(S, K, r, T, \sigma_t^2) \tag{5.10}$$

where O_t is the analytical price of the option pricing function P. The pricing function P typically requires input parameters such as underlying asset price (S), strike price of the option (K), risk free rate (r), time to maturity (T), and volatility of the underlying asset (σ_t^2). The classical option pricing models such as BSOPM simplify the input parameters by assuming no change in the input variables over the life of the option; whereas, more advanced stochastic models such as the GOPM allow the volatility to change according to a GARCH specification. The accuracy of the pricing model is tested using option historical data to back-test all models and benchmark them accordingly.

5.2.3   Delta Hedging

If the correct option pricing model is specified, it should be able to correctly predict the change in the underlying asset. The delta hedge ratio of the option pricing model is calculated as the partial derivative of the analytical option price as follows:

$$\Delta_t = \partial O_t / \partial S_t \tag{5.11}$$

By using the delta of the option pricing model, a portfolio can be replicated periodically to offset any risk associated with the movements in the underlying asset. Therefore, delta hedging can be further used to measure the completeness of the option pricing model. The models are compared and evaluated based on the absolute hedging error.

5.3 Choice of Methodology

Typically, financial models can be grouped into three main categories (Michel Crouhy 2001). The first type is the structural models, which adopt assumptions and simplifications to the underlying processes and market equilibrium which allows equilibrium prices to be derived based on the relationships between underlying asset and their options. For instance, the BSOPM uses the risk-neutral pricing that allows the correct option price to be derived based on the stock value. The second model type is statistic-based models. This type of model makes fewer assumptions about the structure of the market and it depends on the relationship present in the empirical data. Statistical models are used to describe the financial data to give insight into their dynamic behaviour and to observe patterns that can be used for out-of-sample forecasting such as the GARCH model. The third type is a combination of structure and statistical models, which uses the best of both worlds, thereby overcoming some of the other models' limitations and assumptions. The GOPM is of this type since it uses the LRNVR in combination with the GARCH model.

The universal approximation properties (Hornik et al. 1989) of neural networks have contributed to its wide application in different areas. This property allows the neural network to approximate any function of the data allowing the data to drive the model, and eliminating any assumptions about the underlying data. Neural networks are non-linear in nature (Rumelhart et al. 1986), which means that they are capable of capturing the non-linear properties in the data. By using more than one hidden layer, the neural network can partition the sample space and apply different functional forms in each space (Wasserman 1989). This allows the neural network to act as a piece-wise, non-linear model. Neural networks learn by examples, which allow them to capture and infer key aspects and subtle relationships within the data. This generalisation capability allows values to be inferred outside the training set which is particularly useful in the market risk domain. Neural networks are more flexible and adaptable than are the traditional statistical models which have limitations in terms of estimating the relationship between historical and future values of a time series, due to the complexities that exist in real data sets. Also, neural networks can easily adopt new changes in the data (Widrow and Stearns 1985), whereas statistical models would require the complete model to be re-estimated. Neural networks are also adaptable and can be designed to suit any problem; furthermore, extra inputs can be added as required. In this book, we utilise these attributes of the neural networks to address and improve on the limitations found in the literature as highlighted and discussed earlier.

5.4   Research Method

There are two common research methodologies which are social science research and science and engineering-based research which can be adopted in this book. Social science research can be either quantitative or qualitative research. It is often carried out through survey or interview processes. Quantitative research involves extensive data gathering usually using methods such as survey, and statistical analysis of the gathered data in order to prove or disprove various hypotheses that have been formulated. Qualitative research frequently involves in-depth structured or semi-structured interviews that allow one to pursue particular issues of interest that may arise during the interview. It does not normally involve a large sample of data and the information gathered may not be in a form that readily allows statistical analysis. This kind of research can indicate the extent to which the methodology is or is not accepted and sometimes may be able to give the reason for this (Loether 1999; Burstein and Gregor 1999; Nunamaker et al. 1991). However, unlike engineering-based research, this type of research does not explain what a methodology should be and how to produce a new methodology for problem solving. This research only tests or evaluates a method that has already been produced from science and engineering research.

On the other hand, science and engineering-based research is concerned with verifying theoretical predictions or forecast. Galliers (1992) states that in the engineering field, the spirit of 'making something work' is vital and has three levels: conceptual level, perceptual level and the practical level, which are explained below:

- Conceptual level (level one): the creation of new ideas and concepts through analysis.
- Perceptual level (level two): formulating a new method and approach by designing and building the tools or system or environment via implementation.
- Practical level (level three): implementing testing and validation via experimentation using real-world examples.

Science and engineering research can lead to new or improved techniques, architectures, methodologies, or a set of new concepts which all together form a new theoretical framework. Most often, it not only addresses the issues at hand, but also proposes solutions. The objective of this research is to develop a novel framework that applies neural networks technologies to enhance or overcome the limitations of econometric models. To achieve the objective, a framework will be developed which utilises neural networks and proposes solution to the defined objectives. Hence, this research clearly falls into the domain of science and engineering research.

Chapter 6
Volatility Forecasting

In Sects. 2.3 and 4.2, the common volatility modelling oversights that exist in literature were highlighted. In this Chapter, we discuss the potential impact of these oversights on volatility forecasting and provide a methodology for testing the impact of these oversights on the forecasting accuracy of volatility models. In this chapter, we will carry out the experiments needed to evaluate the methodology and we provide an analysis and discussion of the results. The results obtained in this chapter can be used as a guide for developing volatility forecasting solutions. The remainder of this chapter consists of the following. In Sect. 6.3, the volatility models studied in this chapter are explained. The forecast performance evaluation criteria are outlined in Sect. 6.4. The data used in the experiments is analysed in Sect. 6.5. In Sect. 6.6, the experiment approach and methodology are explained. The experimental results are provided in Sect. 6.7. The findings are discussed and concluded in Sect. 6.8.

6.1   Volatility Models

To address the concerns and issues that have been highlighted in Sects. 2.3 and 4.2, in this book, we choose the GARCH, EGARCH and MDN volatility models for investigation. The choice of these models is discussed below. The success of the GARCH model is apparent in the vast body of literature dedicated to this model. The GARCH and EGARCH models will allow us to illustrate the impact of the oversights discussed on the forecasting ability of each of these models, and it will allow us to relate and compare this study to the other work done in this field. Also, this will allow us to demonstrate the importance of selecting the correct volatility model that is most suitable for the return series under study. This is achieved by comparing the performance of the GARCH and EGARCH models across different

© Springer International Publishing AG 2017

F. Mostafa et al., *Computational Intelligence Applications to Option Pricing, Volatility Forecasting and Value at Risk*, Studies in Computational Intelligence 697, DOI 10.1007/978-3-319-51668-4_6

return series. By doing so, we will demonstrate how the wrong model can lead to undesirable results. The performances of the GARCH and EGARCH models are compared and evaluated against the forecasting performance of the neural network. Our choice of neural network is MDN for its ability to model the heteroscedasticity as discussed in Sect. 4.3. MDNs have the same properties as any other neural network; they are flexible and can be designed to suit the task at hand. In this research, we design the MDN model and extend it to become recurrent by feeding back the past period variance or volatility as an input to the model. This will allow the MDN to mimic the GARCH model specification, thereby capturing the key stylised facts of the return series such as volatility clustering.

Below we review the volatility models again to improve the readability of this chapter.

6.1.1   GARCH Model

Bollerslev (1986) extended the conditional variance function of the ARCH model (see Sect. 2.3.7) which he called Generalised ARCH or GARCH. The GARCH model suggests that the conditional variance can be specified as

$$\sigma_t = \alpha_0 + \alpha_1 \varepsilon_{t-1}^2 + \ldots + \alpha_q \varepsilon_{t-q} + \beta_1 \sigma_{t-1} + \ldots + \beta_p \sigma_{t-p} \tag{6.1}$$

$$\alpha_0 > 0$$
$$\text{where, } \alpha_i \geq 0, i = 1, \ldots, q$$
$$\beta_i \geq 0, i = 1, \ldots, p$$

The inequalities are imposed to ensure that the conditional variance is positive. A GARCH process with order p and q is denoted by GARCH (p, q). By expressing Eq. 6.1 as

$$\sigma_t = \alpha_0 + \alpha(B)\varepsilon_t^2 + \beta(B)\sigma_t \tag{6.2}$$

where $\alpha(B) = \alpha_1 B + \ldots + \alpha_q B^q$ and $\beta(B) = \beta_1 B + \ldots + \beta_p B^p$ the variables are polynomials in a backshift operator B. The GARCH model is considered as a generalisation of an ARCH(∞) process, since the conditional variance depends linearly on all previous squared residuals.

6.1.2 EGARCH Model

The GARCH model is well suited for capturing key stylised facts in the return series such as volatility clustering. However, it is not able to capture the leverage effect in return series. This is due to the variance equation being a function of the magnitudes of the lagged residuals and not their signs. Nelson (1991) rectified this issue by proposing the Exponential GARCH (EGARCH). The EGARCH model was formulated with the variance equation that depends on the sign and size of the lagged residual, allowing for the leverage asymmetric effects in the return series.

$$\ln(\sigma_t^2) = \alpha_0 + \sum_{i=1}^{p} \beta_i \ln(\sigma_{t-1}^2) + \sum_{j=1}^{q} \left(\alpha_j \left| \frac{\varepsilon_{t-j}}{\sigma_{t-j}} - \sqrt{\frac{2}{\pi}} \right| + \gamma_j \frac{\varepsilon_{t-j}}{\sigma_{t-j}} \right) \tag{6.3}$$

The presence of leverage effects can be tested by the hypothesis that $\gamma > 0$ and the impact is asymmetric if $\gamma \neq 0$.

6.1.3 Mixture Density Networks

Mixture density networks were first proposed by Bishop (1996), which have been proven to be a very useful instrument in estimating conditional densities with a non-constant variance. The MDN formulation caters for changing variance over the training data set which is a significant advantage over traditional neural networks. The proposed structure of the MDN consists of a single hidden layer with hyperbolic tan activation functions and three output units (priors, centres and widths) per Gaussian. Figure (6.1) illustrates an MDN with two Gaussians.

Fig. 6.1 Recurrent MDN

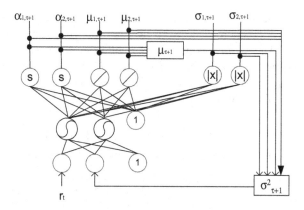

The conditional density of the returns is given by:

$$P(r_t|r_{t-1},\ldots,r_{t-m}) = \sum_{i=1}^{n} \alpha_{i,t} g(\mu_{i,t}, \sigma_{i,t}^2) \tag{6.4}$$

$$g(\mu_{i,t}, \sigma_{i,t}^2) = \frac{1}{\sqrt{2\pi\sigma_{i,t}^2}} \exp\left(-\frac{(r_t - \mu_{i,t})^2}{2\sigma_{i,t}^t}\right) \tag{6.5}$$

where Gaussian parameters μ, α and σ are estimated by

$$\alpha_{i,t} = s(\tilde{\alpha}_{i,t}) = \frac{\exp(\tilde{\alpha}_{i,t})}{\sum_{j=1}^{n}\exp(\tilde{\alpha}_{i,t})} \tag{6.6}$$

$$\tilde{\alpha}_{i,t} = MLP(r_{t-1},\ldots,r_{t-m}), 1 \leq j \leq n \tag{6.7}$$

$$\mu_{i,t} = MLP(r_{t-1},\ldots,r_{t-m}), 1+n \leq j \leq 2n \tag{6.8}$$

$$\sigma_{i,t}^2 = \exp(MLP(r_{t-1},\ldots,r_{t-m})) \tag{6.9}$$

The softmax function $s(\tilde{\alpha}_{i,t})$ ensures that the priors are positive and add up to one. The exponential function in the variance equation ensures that the variance is always positive. The MDN receives a single input and produces three outputs per Gaussian. Typically, when training an MLP on a non-linear regression model, the mean squared error is minimised.

$$MSE = \frac{1}{N}\sum_{t=1}^{N}(y_t - MLP(x_t))^2 \tag{6.10}$$

This method assumes a constant variance and mean, which is the case when modelling homoscedastic data. In the case of heteroscedastic models, a more general error function is needed to cater for the dynamic nature of the underlying data.

The MDN was trained by minimising the negative log likelihood function. By denoting the input data set $D_x = (x_1,\ldots,x_T)$, and the target dataset $D_y = (y_1,\ldots,y_T)$, the full likelihood function can be obtained by multiplying the likelihood of each sample.

$$L = \prod_{t=1}^{T} p(y_t|x_t) \tag{6.11}$$

$$L = \prod_{t=1}^{T}\sum_{j=1}^{M} p_j(x_t)p(y_t|x_t,j) \tag{6.12}$$

$$L = \prod_{t=1}^{T} \sum_{j=1}^{M} p_j(x_t) \frac{1}{\sqrt{2\pi\delta_j^2(x_t)}} \exp\left(\frac{-(y_t - \mu_j(x_t))^2}{2\delta_j^2(x_t)}\right) \qquad (6.13)$$

Since the first product $p_j(x_t)$ in Eq. (6.13) does not depend on the parameters of the MDN, the weights of the network must be optimised with respect to the second product only. Maximising the log likelihood function is the same as minimising the average negative log likelihood function ℓ in (6.14), where ℓ is the error function used for training the MDN.

$$\ell = -\frac{1}{N} \log L \qquad (6.14)$$

$$\ell = -\frac{1}{N} \sum_{t=1}^{N} \log P(y_t|x_t) \qquad (6.15)$$

Let us consider an MDN with one Gaussian; the likelihood function would be:

$$\ell = \frac{1}{N} \sum_{t=1}^{N} \left(\frac{1}{2} \log(2\pi\delta_t^2) + \frac{(y_t - \mu_t)^2}{2\delta_t^2}\right) \qquad (6.16)$$

It can be seen that the likelihood function is a special case of the *MSE* in Eq. (6.10), given a single Gaussian and a constant variance. Hence, the log likelihood function can be used on the out-of-sample dataset for model evaluation. The MDN was trained by minimising the function ℓ with the scale conjugate descent algorithm. The log likelihood function ℓ will be referred to as the 'loss function', which in turn would be used to evaluate the model for in-sample and out-of-sample performance.

6.1.3.1   Recurrent Mixture Density Networks

Given the recurrent nature of the GARCH model, the MDN is extended to include the lagged variance as an input parameter. As noted by Schittenkopf et al. (1998), if the input to the MDN is extended to include lagged values of the variance δ_{t-1}^2, the networks become a generalisation of the GARCH model. With the generalisation capability of the MDN to approximate the distribution of the underlying data, the MDN should yield better forecasts as observed by Ormoneit and Neuneier (1996) and Schittenkopf et al. (1998). The recurrent mixture density network (rMDN) can approximate the distribution of the underlying data, thereby overcoming the assumption of normality in the GARCH Model. In Cooperation et al. (1999), the rMDN was constructed so that the output of the Gaussians is designed with its own hidden units. In our approach, the rMDN is fully interconnected.

Fig. 6.2 Stock price time
series

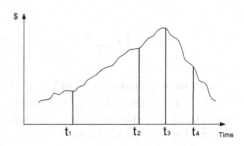

To demonstrate the importance of selecting the right model for a given time
series, different rMDN structures are investigated. The rMDN structures were
varied by changing the hidden units and the number of Gaussians. As shown in
Fig. 6.2, the conditional variance of the previous period (σ_{t-1}^2) is fed back as an
input at time t which is in line with the GARCH model. The calculation of the
conditional mean and conditional variance are as follows:

$$\mu_{t+1} = \sum_{i=1}^{n} \alpha_{i,t+1} \mu_{i,t+1} \tag{6.17}$$

$$\sigma_{t+1}^2 = \sum_{i=1}^{n} \alpha_{i,t+1} \left(\delta_{i,t+1}^2 + (\mu_{i,t+1} - \mu_{t+1})^2 \right) \tag{6.18}$$

Another attractive feature of the rMDN model is its ability to model
time-dependent higher order moments, such as skewness and kurtosis, in contrast to
a constant skewness and kurtosis by the GARCH models as demonstrated by
Schittenkopf et al. (2000). The skewness and conditional kurtosis can be calculated
as follows:

$$s_{t+1} = \frac{1}{\delta_{t+1}^3} \sum_{i=1}^{n} \alpha_{i,t+1} \left(3\delta_{i,t+1}^2 (\mu_{i,t+1} - \mu_{t+1}) + (\mu_{i,t+1} - \mu_{t+1})^3 \right) \tag{6.19}$$

$$k_{t+1} = \frac{1}{\delta_{t+1}^4} \sum_{i=1}^{n} \alpha_{i,t+1} \left(3\delta_{i,t+1}^4 + 6\delta_{i,t+1}^2 (\mu_{i,t+1} - \mu_{t+1})^2 + (\mu_{i,t+1} - \mu_{t+1})^4 \right)$$

$$\tag{6.20}$$

Below is an illustration of the RMDN with two hidden units and two Gaussians.

6.2 Issues Investigated

In Sect. 4.2, we provide a detailed review of the literature pertaining to volatility modelling using neural networks and MDNs in particular. The majority of the literature on neural networks for volatility forecasting, consists of hybrid techniques which involve combining neural networks with other time series models. This approach has been adopted due to the inability of the MLP to model heteroscedasticity in the return series. This weakness has been overcome by the MDN formulation which allows the variance to change in the training set. Hence, we focus our attention on MDNs since they are most suitable for forecasting volatility. There are a handful of papers that attempt to forecast volatility using MDNs. An analysis of these papers reveals an inconsistency of approach which is common across neural network forecasting literature. This is especially so when selecting the correct MDN structure and selecting the optimal training data set for optimising the model. Table 6.1 contains the summary of all research conducted thus far. Table 6.1 also shows the inconsistencies in the approach when building and modelling the MDN. As discussed in Sect. 4.1, selecting the number of hidden units without verification can lead to issues such as over-fitting of the training data set. Therefore, by having a large number of parameters in the model such as too many hidden units, this would cause the model to memorise the training data set. This will cause the model to have poor out-of-sample forecasting performance due to the loss of generalisation capabilities of the model.

This issue is more problematic in neural network models as there is no formal method for selecting the correct model based on the underlying data. In Sect. 4.1, we also explain the issues of selecting the training data set in an ad hoc manner and the adverse effect it has on the models forecasting performance. For instance, if a large data set is used to optimise the model, we run a risk of having old behaviours in the past that are no longer relevant. So the model forecast could be inaccurate and misleading. Also, using less data would cause instability in the model's performance caused by the model parameters not having the optimal values. The impact of these common oversights and inconsistencies in modelling volatility is apparent in the results obtained in the research. This can be seen specifically in

Table 6.1 Research papers, MDN in volatility forecasting

Paper	Number of observation used in training	Model structure
Nabney and Cheng (1997)	2505	• 1–5 Gaussians • Up to 30 hidden units • Non-recurrent MDN
Schittenkopf et al. (2000)	900	• 2 Gaussians • 3 hidden units • Recurrent
Miazhynskaia et al. (2003)	500	• 1–2 Gaussian • Recurrent

Miazhynskaia et al. (2003) as the results varied between models and across different time series. In this book, we illustrate and study the impact of these oversights by showing the forecasting performance impact that they have on all models and demonstrate superior forecasting accuracy of the MDN subject to selecting the correct model structure. We will demonstrate the impact of the following issue on the model's forecasting performance:

- Selecting the wrong training data set
- Selecting the wrong number of hidden units
- Selecting the wrong number of Gaussian
- Using the wrong time series model.

The results of this experiment will be useful for model specification and design in accord with the return series.

6.3   Solution Overview

The introduction of the ARCH models was a major breakthrough in volatility forecasting, where for the first time, key stylised facts of the return series can be captured in a time series model. This has contributed to vast interest in the ARCH models evidenced by the large body of literature available today. The ARCH model formulation was adequate for capturing key stylised facts like volatility clustering and persistence. However, other stylised facts such as leverage effects were not catered for. For this reason, many variations of the ARCH model have been developed with the aim of capturing different attributes of the returns series. The main drawback of all volatility models is the assumption of normality in the return series. While this assumption is made to simplify the formulation of the model, at the same time it significantly undermines the forecasting performance since the distribution of the return time series is known to have fat tails. This assumption has been relaxed by allowing the GARCH process to follow a different distribution such as *t-student* distribution. The *t-student* has fatter tails and is much suitable for the return series; however, this is still forcing the returns to conform to a specific distribution.

The drawbacks and limitations of the volatility models have motivated researchers to create alternative methods to overcome these weaknesses. Such new methods include the adoption of neural networks in volatility forecasting. Just like the ARCH models, the neural networks have endured varying success in the literature on volatility forecasting. The majority of neural network research applies a hybrid approach to volatility forecasting where the neural network is combined with time series models to produce a volatility forecast. This approach seems to have more success than the method of using neural networks to forecast conditional volatility. The limited success of neural networks in forecasting conditional volatility can be explained by the inability of the neural network to capture the heteroscedastic property of the return series as discussed in Sect. 4.2. This explains why the majority of the research has adopted a hybrid approach by combining neural networks and other time series models since a stand-alone neural network would be ineffective. The MDN formulation overcomes the limitation of the neural networks by allowing the variance to be functions of the input value (see Sect. 4.3). By changing the variance over the entire training population, this allows the MDN to model time varying variances which is ideal for volatility forecasting.

Even though the MDN is suitable for modelling heteroscedastic data, results presented in the literature are still inconsistent. Also, the MDN forecasting performance seems to be on par with that of the GARCH model. This is somewhat surprising given the features of the MDN and its modelling capabilities. The MDN has a high number of variables that complicates the model building exercise. In the literature, the number of hidden units is chosen at random. The choice of the number of hidden units can greatly impact on the outcome of the model. Increasing the number of hidden units also increases the chance of over-fitting of the data; at the same time, the use of fewer hidden units would lead to under-fitting and potentially cause wrong forecasts. Since the MDN can approximate any distribution as the number of Gaussians approaches infinity, it is desirable to have a large number of Gaussians. However, as the number of Gaussians increases, the number of weights in the output layer also increases by a factor of three. This increases the number of weights in the network which can also result in over-fitting of the data. In the literature, the common approach is to choose the number of hidden units at random without any verification, which will certainly have an impact on the model's stability and performance.

Model optimising involves searching for the best fit for the model variables that produces the lowest error for a given training data set. To produce stable and consistent results, the training data set must be chosen appropriately. In Chap. 4, the literature review reveals the length of the training set chosen in an ad hoc manner. This is a common oversight where in most cases the in-sample is chosen to be 1000 observations without any validation. The impact of this oversight can be explained as follows:

If a model is optimised using observations from t_1 to t_2 (see Fig. 6.2), the model parameters will reflect the information and behaviours available between periods t_1 and t_2. Forecasting values in the period between t_2 and t_3 can be achieved with a certain degree of accuracy. Since there is no break or fundamental shifts in the data generating process. However, if the model is optimised using data from t_1 to t_3, accurately forecasting values between period t_3 and t_4 would be very difficult to achieve. The in-sample data set used to optimise the model did not include such structure breaks or turning points; therefore, the model would be unable to predict values in that region. This illustrates the importance of having a meaningful training data set that is a true representation of the time series. Another issue in choosing the training data set is the number of observations to be used in optimising the model. That is, by optimising the model using a large in-sample data set, this can lead to over-fitting, and using a small data set can lead to under fitting. In both cases, the model forecasting performance is affected. These critical issues have not been dealt with in the ARCH literature which can potentially explain the discrepancies of the reported model results.

The abovementioned oversights in the literature have contributed to the mixed results that have been reported. To demonstrate the effect of these oversights, we conduct a thorough experiment to address the following questions:

- How does the number of observations in the training data set impact on the GARCH and MDN models' forecasting performance?
- How do the number of hidden units and number of Gaussians impacts the forecasting performance of the MDN?
- How does the number of Gaussians in the MDN impact on the forecasting performance of the MDN?
- How do the forecasting models performance vary across different time series?
- Can MDN outperform the GARCH model?

To address these concerns, we select the GARCH and EGARCH models to be studied in this book. Selecting both models allows us to benchmark the GARCH model against the EGARCH to demonstrate the effect of choosing one volatility model against the other. We also choose MDN as it has been demonstrated in the literature to be the most suitable neural network model for forecasting volatility. The experiment includes testing all models on four different time series. Each model will be optimised using 750 and 1000 observations. For the MDN models, we use different combinations for the hidden units and number of Gaussians. The hidden units range from 1 to 5 and Gaussian from 2 to 3 Gaussians. The inputs to the MDN are designed to mimic the GARCH model which included the previous volatility value and return. The model's evaluation is based on 1, 10 and 30 day forecasts. The model's performances are benchmarked according to three main criteria: loss function, normalised mean absolute error, and hit rate. This approach will demonstrate the impact of the common oversights in literature on the forecasting performance of the model. This approach will also demonstrate and explain the mixed results reported by many authors. Given the capabilities and features of MDN as discussed in Chap. 4, Sect. 4.2, the MDN should have superior forecasting performance to other time series models given the optimal model is selected in line with the concerns highlighted in this section. By addressing these concerns, we make several contributions by demonstrating:

- The effect of using the wrong number of observations in the training data set on the out-of-sample forecasting capabilities of the GARCH and MDN models
- The forecast performance when using the same model across different time series
- How the number of hidden units impacts on the forecast performance of the MDN model
- How the number of Gaussians impacts on the performance of the MDN
- The superiority of the MDN over GARCH models subject to selecting the correct model specification.

6.4 Forecast Evaluation

Evaluating volatility forecasts is problematic because the volatility is a latent variable which is not directly observable. If the return innovation is given by $r_t = \sigma_t \varepsilon_t$, where ε_t random variable drawn from $N(0, 1)$, and σ is the variance, then

$$E_{t-1}[r_t^2] = E_{t-1}[\sigma_t^2 . \varepsilon_t^2] = \sigma_t^2 \tag{6.21}$$

This seems to justify the use of the return squared innovation as a proxy for the true volatility. The use of return squared innovation as a proxy for the true volatility results in a very noisy measurement due to the idiosyncratic error term (ε_t^2). Andersen and Bollerslev (1997) stated "Consequently, the poor predictive power of volatility models, when judged by standard forecast criteria using r_t^2 as a measure for ex-post volatility, is an inevitable consequence of the inherent noise in the return generating process". Lopez (2001) has also shown that this method tends to be 50% greater or smaller than the true volatility 75% of the time. Andersen and Bollerslev (1997) have demonstrated that the high frequency return series provides a more accurate volatility measurement. Since our data consists of daily stock prices, r_t^2 is used as a proxy for the true volatility.

The performance of the volatility models are measured by the loss function ℓ, normalised mean absolute error (NMAE) and hit rate (HR).

$$NMAE = \frac{\sum_{t=1}^{N} |r_{t+1}^2 - \hat{\delta}_{t+1}^2|}{\sum_{t=1}^{N} |r_{t+1}^2 - r_t^2|} \tag{6.22}$$

$$HR = \frac{1}{N} \sum_{t=1}^{N} \theta_t$$

$$\text{where, } \theta_t = \begin{cases} 1 : (\delta_{t+1}^2 - r_t^2)(r_{t+1}^2 - r_t^2) \geq 0 \\ 0 : else \end{cases} \tag{6.23}$$

The NMAE measures the mean absolute error of the volatility model, compared to the true volatility r_t^2, which should be smaller than the naive model $\delta_t^2 = r_t^2$. This measure is the least reliable of the three measuring criteria used. Since it compares the forecasted volatility with the returns squared, this has some major drawbacks as stated previously. The hit rate (HR) is the measure of frequency of the correctly predicted increase (or decrease) in volatility. An HR value of 0.5 indicates a forecast that is no better than a random predictor of increase (or decrease) in volatility. This measure is a better indicator than the NMAE since it does not rely on the magnitude of the squared return; only that the predicted direction is taken into account. Since the loss function can be calculated for in-sample and out-of-sample data sets, it can be used as an error measurement for in-sample and out-of-sample tests. The minimisation of the loss function for an MDN with one

Gaussian is simply reduced to the MSE of the model. The calculation of this measure is independent of the return squared; therefore, it is considered to be the most accurate performance measure of the model. The model performance would be ranked according to the smallest value for the loss function with the heights HR and smallest NMAE respectively.

6.5   Data Analysis

The data used in this research consists of daily close prices for four companies listed on the Australian Stock Exchange (ASX). The companies used are BHP BLT (BHP), QANTAS (QAN), National Australia Bank (NAB) and ANZ Bank (ANZ). The in-sample data consists of two sets, 1000 and 750 returns series. The consecutive 100 returns are used for out-of-sample testing. The daily close prices were transformed to daily returns through Eq. 6.24.

$$r_t = \log\left(\frac{close\_price_{t+1}}{close\_price_t}\right) \tag{6.24}$$

Below is a sample of the ANZ return series. The existence of volatility clustering in return (Fig. 6.3).

We use two different lengths for the in-sample data. The sample length is chosen to be 1000 and 750 returns. Below are some statistics of the data sets (Tables 6.2 and 6.3).

The BHP return series has a kurtosis greater than three, which means the distribution is leptokurtic, whereas all other series are platykurtic. All return series have a long left tail, due to the skewness being less than zero, whereas QAN (for 1000 returns) have a long right tail.

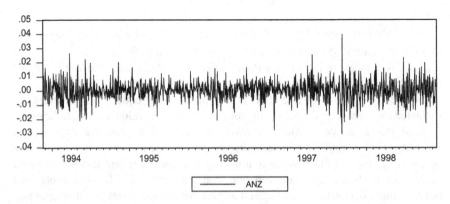

Fig. 6.3 Sample of daily returns series for ANZ

Table 6.2 Basic statistics for 1000 returns

	BHP	QAN	NAB	ANZ
Mean	−0.0001	0.0008	0.0006	0.0008
Standard Deviation	0.0124	0.0169	0.0116	0.0151
Skew	−1.3415	0.0726	−0.4693	−0.0054
Kurtosis	14.6319	1.8136	2.1956	2.6929

Table 6.3 Basic statistics for 750 returns

	BHP	QAN	NAB	ANZ
Mean	0.0000	0.0002	0.0001	0.0006
Standard Deviation	0.0124	0.0161	0.0108	0.0142
Skew	−1.6521	−0.2625	−0.6760	−0.0118
Kurtosis	18.9723	1.8277	2.1545	1.1603

6.6 Experimentation

In this book, we benchmark the volatility forecast performance of the MDN, GARCH and EGARCH models. The experiment is carried out in line with the methodology outlined in Chap. 4. All models are optimised using 750 and 1000 returns. However, for the MDN we vary the hidden units from 1 to 5 and the number of Gaussians from 2 to 3. More Gaussians are desirable however, as discussed in Chap. 4, since by increasing the number of Gaussians, this would increase the number of hidden weights by a factor of three. This can lead to over-fitting of the data. The GARCH and EGARCH models were optimised by maximising the log likelihood with the BHHH (Berndt-Hall-Hall-Hausman) algorithm. However, the MDN is optimised by minimising the negative log likelihood with the scale conjugate descent algorithm (using the Netlab package with Matlab).

Once the models have been optimised, we first conduct in-sample testing. That is after training the model, 100 points from the in-sample data set is presented back to the model, this will allow us to check whether the model is able to reproduce the trained data. Failure to successfully reproduce data from the training set is an indication that the model has not been properly optimised. Once we have verified the models via in-sample testing, we perform out-of-sample testing. This is carried out for 1, 10 and 30 day forecasts. This will allow us to verify how the models measure up in terms of short-, medium- and long-term forecasting.

6.7 Results

In this section, we provide the in-sample and out-of-sample results for all experiments.

6.7.1   In-Sample Testing

All GARCH models are stationary with high persistence ($\omega + \beta < 1$) for both 1000
and 750 in-sample series except for BHP. The parameters for the EGARCH model
show asymmetric and leverage effect ($\gamma \neq 0$ and $\gamma < 0$) for all companies except for
BHP, which exhibits only asymmetric effects ($\gamma \neq 0$). Tables 6.4, 6.5, 6.6 and 6.7
show the parameter estimates for the GARCH and EGARCH models for 1000 and
750 returns (Tables 6.8 and 6.9).

As discussed in Sect. 6.4, the loss function is a better measurement of how well
the model fits the data. In all cases, the GARCH and EGARCH models give similar
results for all error measurements. For optimal performance of the MDN, the

Table 6.4 GARCH
parameters—750 returns

	ANZ	NAB	BHP	QAN
μ	0.00113	0.00012	3.25000	0.00011
α	0.00001	0.00001	0.00009	0.00001
β	**0.079**	**−0.017**	**0.326**	**0.081**
ω	**0.875**	**0.972**	**0.153**	**0.897**

Table 6.5 GARCH
parameters—1000 returns

	ANZ	NAB	BHP	QAN
μ	0.00129	0.00050	−0.00007	0.00055
α	0.00004	0.00002	0.00010	0.00001
β	**0.197**	**0.053**	**0.261**	**0.082**
ω	**0.648**	**0.825**	**0.085**	**0.896**

Table 6.6 EGARCH
parameters—750 returns

	ANZ	NAB	BHP	QAN
μ	0.00036	0.00006	0.00002	−0.00007
ω	−0.60559	−3.65301	−2.59297	−0.33302
α	0.138	−0.119	0.092	0.155
γ	**−0.054**	**−0.034**	**0.017**	**−0.016**
β	0.952	0.654	0.764	0.979

Table 6.7 EGARCH
parameters—1000 returns

	ANZ	NAB	BHP	QAN
μ	0.00092	0.00054	−0.00019	0.00003
ω	−1.86498	−1.56775	−2.59297	−0.98046
α	0.303	0.117	0.092	0.229
γ	**−0.116**	**−0.010**	**0.017**	**−0.026**
β	0.840	0.860	0.764	0.919

Table 6.8 GARCH and EGARCH in-sample results for 1000 data points

	BHP		Qan		NAB		ANZ	
	NMAE	HR	NMAE	HR	NMAE	HR	NMAE	HR
GARCH	0.922	72	0.843	74	0.921	70	0.919	74
EGARCH	0.930	73	0.820	80	0.920	70	0.903	72
LF GARCH	2.999		2.712		3.051		2.836	
LF EGARCH	3.002		2.712		3.048		2.845	

Table 6.9 GARCH and EGARCH in-sample results for 750 data points

	BHP		Qan		NAB		ANZ	
	NMAE	HR	NMAE	HR	NMAE	HR	NMAE	HR
GARCH	0.925	71	0.839	74	0.930	72	0.899	78
EGARCH	0.954	72	0.839	75	0.914	72	0.875	77
LF GARCH	2.256		2.082		2.336		2.154	
LF EGARCH	2.260		2.082		2.341		2.158	

structure is chosen to have 1–5 hidden units, with 2 and 3 Gaussians. Tables 6.10 and 6.11 contain the MDN in-sample results for 1000 and 750 in-sample series. It will be unfair to compare the in-sample statistics of the GARCH and EGARCH with those of the MDN. The reason for this is that the GARCH and EGARCH models were fitted with different BHHH (Berndt-Hall-Hall-Hausman) algorithms by maximising the log likelihood function, (using the eviews package) whereas, the MDN was trained by minimising the log likelihood with the scale conjugate descent algorithm (using the Netlab package with Matlab). Our main interest is in the out-of-sample performance of the models. The numbers shown in bold and underline in the tables below highlight the best performing model.

6.7.2   Out-of-Sample Forecast Performance

The objective of this research is to compare the forecast performance of the MDN network with that of the GARCH and EGARCH models with the emphasis on the right model selection for the given time series. The models are compared based on 1, 10 and 30 steps ahead forecast. The model forecast is conditional on the information set given at time t. In the case of the long-term forecast, the information set becomes less significant as the forecast horizon increases. Below we summarise the results of the best performing models. The full results are reported in Appendix A.

Table 6.10 In-sample statistics for 1000 returns

	BHP			ANZ			NAB			QAN		
	NMAE	HR	LLF	NMAE	HR	LLF	NMAE	HR	LLF	NMAE	HR	LLF
2G2H1k	0.952	0.71	2.998	0.906	0.74	2.817	0.941	0.710	3.079	0.796	0.75	2.689
2G3H1k	*0.883*	*0.73*	*3.000*	0.902	0.74	2.817	0.950	0.71	3.075	0.796	0.73	2.690
2G4H1k	0.966	0.71	3.003	0.908	0.74	2.816	0.953	0.71	3.072	*0.786*	*0.76*	*2.689*
2G5H1k	0.912	0.73	2.997	0.957	0.72	2.809	0.946	0.71	3.075	0.795	0.75	2.690
2G1H1k	0.885	0.73	2.953	0.907	0.74	2.817	0.949	0.71	3.078	0.796	0.73	2.689
3G2H1k	1.248	0.7	2.953	0.887	0.74	2.844	0.936	0.71	3.086	0.797	0.76	2.694
3G3H1k	1.282	0.69	2.947	0.907	0.74	2.836	*0.881*	*0.71*	*3.091*	0.796	0.76	2.693
3G4H1k	0.998	0.72	2.993	0.917	0.71	2.840	0.909	0.71	3.091	0.797	0.76	2.694
3G5H1k	1.268	0.7	2.949	*0.884*	*0.74*	*2.841*	0.912	0.71	3.091	0.796	0.73	2.694
3G1H1k	0.990	0.72	2.994	0.888	0.74	2.843	0.940	0.71	3.086	0.797	0.74	2.692

Table 6.11 In-sample statistics for 750 returns

	BHP			ANZ			NAB			QAN		
	NMAE	HR	LLF	NMAE	HR	LLF	NMAE	HR	LLF	NMAE	HR	LLF
2G2H750	1.052	0.71	2.240	0.870	0.75	2.148	0.897	0.720	2.354	0.795	0.77	2.047
2G3H750	0.902	0.73	2.255	0.908	0.77	2.11	0.898	0.71	2.363	0.795	0.77	2.048
2G4H750	0.999	0.72	2.247	0.917	0.76	2.139	0.891	0.72	2.355	0.794	0.77	2.047
2G5H750	0.949	0.71	2.253	0.867	0.75	2.148	0.893	0.72	2.352	0.794	0.77	2.048
2G1H750	*__0.872__*	*0.73*	__2.255__	0.927	0.75	2.140	0.906	0.71	2.354	0.793	0.77	2.047
3G2H750	1.204	0.7	2.219	0.860	0.77	2.158	__0.885__	*0.72*	__2.368__	0.794	0.74	2.053
3G3H750	0.905	0.73	2.255	0.861	0.75	2.156	0.891	0.72	2.360	0.796	0.78	2.052
3G4H750	1.209	0.7	2.218	*0.859*	*0.77*	*2.159*	0.882	0.72	2.367	0.754	0.76	2.074
3G5H750	1.223	0.7	2.217	0.867	0.76	2.155	0.884	0.72	2.367	*0.753*	*0.76*	*__2.074__*
3G1H750	1.205	0.7	2.219	0.861	0.76	2.156	0.892	0.72	2.366	0.795	0.77	2.053

6.7.3   One-Day Volatility Forecast

For the one-day-ahead forecast, we perform 100 one-day forecasts. This is achieved by training the model using returns from period 1 to T, then forecasting volatility at time T + 1. The model is retrained using returns from period 2 to T + 1 and then forecasting the volatility at period T + 2.

In Table 6.12, QAN and NAB provide a much better performance using 750 compared to 1000 returns. BHP and ANZ, on the other hand, gave poorer results for 750 returns and much better results for 1000 returns (see Table 6.13). The EGARCH model tends to be a better choice for QAN, NAB and ANZ, whereas the GARCH model seems to be a better choice for BHP. These findings are consistent with the results reported in the data analysis in Sect. 6.4 (Table 6.14).

It is important to note the optimal MDN structure is different across time series. For example, in Table 6.15, the optimal structure for BHP is 3 Gaussians and 2 hidden units, whereas for ANZ, it is 3 Gaussians and 4 hidden units. So it is very important to optimise the structure of the MDN with respect to the time series understudy. For all time-series (except QAN), the MDNs seem to perform much better for 1000 returns, whereas, QAN performed better with 750 returns. Nevertheless, the MDN models demonstrated superior performance to that of GARCH and EGARCH models.

Table 6.12 1 day forecast for 1000 returns

	Model	NMAE	HR	ℓ
BHP	EGARCH	0.667	0.74	2.19
ANZ	GARCH	0.796	0.73	2.578
NAB	EGARCH	0.745	0.74	2.754
QAN	GARCH	0.914	0.68	2.379

Table 6.13 1 day forecast for 750 returns

	Model	NMAE	HR	ℓ
BHP	EGARCH	0.847	0.65	3.007
ANZ	GARCH	0.705	0.74	2.836
NAB	EGARCH	0.705	0.74	2.48
QAN	GARCH	0.746	0.72	2.52

Table 6.14 MDN 1 day forecast for 1000 returns

	rMDN	NMAE	HR	ℓ
BHP	3G2H1k	0.662	0.74	2.261
ANZ	3G4H1k	0.753	0.75	1.282
NAB	3G2H1k	0.72	0.75	1.34
QAN	3G4H1k	0.991	0.64	1.342

Table 6.15 MDN 1 day forecast 750 returns		rMDN	NMAE	HR	ℓ
	BHP	3G5H750	0.742	0.72	3.906
	ANZ	3G5H750	0.857	0.65	1.369
	NAB	3G2H750	0.722	0.73	1.34
	QAN	3G5H750	0.738	0.76	1.25

6.7.4 10 and 30 Days Forecast

The forecast performance of the GARCH and EGARCH models for 10- and 30-days-ahead forecasts are shown in Tables 6.16, 6.17 and 6.18. All models seem to perform better for the 10-day forecast with 1000 returns, except for the QAN, which shows slightly better performance with 750 returns. Whereas, for the 30-day forecast, BHP and ANZ do much better with 1000 returns and QAN and NAB perform much better with 750 returns. EGARCH outperformed the GARCH model on all tests (Table 6.19). The numbers shown in bold in the tables below indicate the most performant model.

Long-term forecasts with the MDN are not as easily obtained. The returns r_{t+1}, are not directly observed. Hence, we need to simulate the future returns using the Monte Carlo simulation as described below.

- Draw 15,000 normal distributed random numbers according to N(0,1)
- For each of the numbers generated, multiply it by the standard deviation of the returns at time $t+1$ to obtain the disturbance for $t+1$ according to $\varepsilon_{t+1} \approx \sigma_{t+1}N(0,1)$

Table 6.16 10-day forecast for GARCH and EGARCH models using 1000 days			NMAE	LLF	HR
	BHP	GARCH	*0.960*	*0.727*	*0.600*
		EGARCH	0.969	0.707	0.600
	NAB	GARCH	*0.606*	*2.529*	*0.800*
		EGARCH	0.593	2.437	0.800
	QAN	GARCH	*0.702*	*2.726*	*0.600*
		EGARCH	0.710	2.723	0.600
	ANZ	GARCH	0.498	2.277	0.800
		EGARCH	*0.524*	*2.103*	*0.900*

Table 6.17 10-day forecast for GARCH and EGARCH models using 750 days			NMAE	LLF	HR
	BHP	GARCH	0.729	2.746	0.500
		EGARCH	*0.755*	*2.889*	*0.600*
	NAB	GARCH	0.793	3.222	0.800
		EGARCH	*0.777*	*3.218*	*0.800*
	QAN	GARCH	2.943	2.544	0.600
		EGARCH	*2.506*	*2.549*	*0.700*
	ANZ	GARCH	*0.724*	*3.027*	*0.600*
		EGARCH	0.752	3.022	0.600

Table 6.18 30-day forecast for GARCH and EGARCH models using 1000 days

		NMAE	LLF	HR
BHP	GARCH	0.753	0.822	0.767
	EGARCH	*0.758*	*0.598*	*0.767*
NAB	GARCH	*0.693*	*2.564*	*0.833*
	EGARCH	0.687	2.528	0.833
QAN	GARCH	*0.798*	*2.811*	*0.700*
	EGARCH	0.812	2.802	0.700
ANZ	GARCH	0.730	2.359	0.767
	EGARCH	*0.741*	*2.264*	*0.800*

Table 6.19 30-day forecast for GARCH and EGARCH models using 750 days

		NMAE	LLF	HR
BHP	GARCH	0.772	2.814	0.633
	EGARCH	*0.743*	*2.721*	*0.667*
NAB	GARCH	*0.705*	*2.273*	*0.767*
	EGARCH	0.705	2.206	0.767
QAN	GARCH	1.018	2.436	0.667
	EGARCH	*0.950*	*2.426*	*0.700*
ANZ	GARCH	*0.772*	*2.814*	*0.600*
	EGARCH	0.743	2.721	0.600

- Calculate $r_{t+1} = \mu_{t+1} + \varepsilon_{t+1}$ by adding the disturbance ε_{t+1}, to the conditional mean μ_{t+1} (as calculated at time t + 1)
- Feed the simulated returns and σ_t^2 as inputs to the MDN to obtain the mean $\hat{\mu}_{t+1}$ and variance $\hat{\sigma}_{t+1}^2$ (using Eqs. 6.17 and 6.18)
- The expected variance and mean for t + 1 is the average value of each
- Repeat the above steps by using $\hat{\sigma}_{t+1}^2$, for the next forecast.

For all volatility models, the accumulated conditional variance for the day after tomorrow is calculated as the average conditional variance, as shown below:

$$\hat{\sigma}_{t+1}^2 = \frac{1}{T-1} \sum_{i=2}^{T} E(\sigma_{t+i}^2 | I_t) \tag{6.25}$$

QAN and BHP gave acceptable forecasts for 10 days with 750 returns. ANZ and NAB companies perform better with 1000 returns. For the 30 days forecast, only QAN gave good performance using 750 returns. Interestingly, the same MDN structure that performed well for the 10-day forecast gave a completely different result for the 30 day forecast. Also, the EGARCH model for the QAN (with 750 returns), outperforms the MDN on the ℓ measure. But the MDN out performs the EGARCH on the other two measures, HR and NMAE. The NMAE for the EGARCH gives a value which is greater than 1. EGARCH for NAB with 750

Table 6.20 rMDN 30-day results using 1000 returns

	rMDN	NMAE	HR	ℓ
BHP	2G2H1k	0.735	0.83	1.366
ANZ	2G4H1k	0.750	0.80	2.260
NAB	3G2H1k	0.676	0.83	2.384
QAN	3G3H1k	0.845	0.60	2.855

Table 6.21 rMDN 30-day results using 750 returns

	rMDN	NMAE	HR	ℓ
BHP	3G1H750	0.740	0.67	3.575
ANZ	3G3H750	0.752	0.63	3.369
NAB	2G8H750	0.704	0.77	3.633
QAN	2G2H750	0.597	0.80	2.365

Table 6.22 rMDN 10-day results using 750 returns

	rMDN	NMAE	HR	ℓ
BHP	2G5H750	0.719	0.6	3.341
ANZ	3G3H750	0.797	0.6	3.371
NAB	2G8H750	0.774	0.8	3.633
QAN	2G5H750	0.585	0.8	3.605

Table 6.23 rMDN 10-day results using 1000 returns

	rMDN	NMAE	HR	ℓ
BHP	2G2H1k	0.919	0.7	1.297
ANZ	2G8H1k	0.535	0.9	2.095
NAB	3G2H1k	0.571	0.8	2.051
QAN	3G3H1k	0.681	0.8	3.018

returns performs slightly better than MDN on LLF measure, but performs exactly the same on the HR and NMAE measures (Tables 6.20, 6.21, 6.22 and 6.23).

6.8 Conclusion

This study has demonstrated the ability of the MDN to model complex time series data. The superior performance of the MDN relative to traditional GARCH models is attributed to the flexibility and the dynamic nature of the MDN. Given the success of the MDN, there is still room for improvements, such as pruning the neural network and including other exogenous variables that have some explanatory power. The emphasis is mainly on the right model selection. The selection of the correct model and the use of an optimal training data set go hand-in-hand to provide reliable and stable results. This was demonstrated by varying the number of Gaussians, hidden units, and varying the number of observations in the in-sample lengths. In addition, we tested the models using four different time series. The performance of the models varied dramatically with these factors, which explains why other researchers show poor results when comparing MDNs to other traditional models without optimising their model. In this chapter, we have made the following contributions:

- Demonstrated the effect of using different numbers of observations in the in-sample set on the model out-of-sample forecasting accuracy
- Shown how volatility models' performances vary across time series; therefore, special attention must be given when selecting the correct model based on the statistical properties of the return series
- Demonstrated choice of the wrong number of hidden units by varying the number of hidden units
- MDN models with the same hidden unit units and Gaussians performed differently across different time series
- Results demonstrating the superiority of the MDN over GARCH models where the correct MDN model is specified.

Chapter 7
Option Pricing

In Sect. 7.1, we review several methods for option pricing in research. Specifically, in Sect. 7.1.5, we review Neural Net Methods for options pricing; the strengths and weaknesses of each of the applied methods is discussed. This review revealed a trend in the methodology applied in this domain which consists of training the neural networks on the option price. This requires the neural network to provide a mapping from input space to the option price. Also, the majority of research benchmarked the neural network pricing capabilities against the BSOPM and its variants. The results demonstrate the superior pricing capabilities over the BSOPM. However, there is no evidence to show how the neural network models would fare against more sophisticated option pricing models such as the GOPM. Our analysis reveals that the reason behind this trend is the inferior capabilities of neural networks to price options out-of-sample with high accuracy relative to advanced option pricing models. In Sects. 7.1.5 and 7.2, we explain the complexities behind option pricing and provide justification for the weakness and drawbacks in the neural network approach adopted in literature. In Sect. 7.3 we propose a solution to this problem which overcomes the limitation and inefficiencies of the previous methods. To further evaluate our proposed method, a delta hedging scenario is examined which will allow us to evaluate the completeness of the model. The remainder of this chapter consists of the following: in Sect. 7.2 the issues investigated in this study are highlighted; in Sect. 7.3, an explanation is given of the option pricing model developed in this study. The data used in this study is analysed in Sect. 7.4. The experiment design is discussed in Sect. 7.5 and performance measures explained in Sect. 7.6. The delta hedging exercise is explained in 7.7. The results are presented in Sect. 7.8 and a discussion is given in Sect. 7.9. In Sect. 7.10, the empirical dynamics of the volatility smile is analysed. The summary and conclusion is provided in Sect. 7.11.

© Springer International Publishing AG 2017
F. Mostafa et al., *Computational Intelligence Applications to Option Pricing,
Volatility Forecasting and Value at Risk*, Studies in Computational
Intelligence 697, DOI 10.1007/978-3-319-51668-4_7

7.1 Option Pricing Models

Since the publication of the BSOPM in 1973 (Black and Scholes 1973), it remains the most quoted scientific paper in the world. The model made a key contribution to option trading, allowing investors to calculate a fair value for an option contract. This model had its limitations, which stem from the unrealistic assumptions used in the derivation of the BSOPM. The BSOPM's behaviour and limitations have been well documented in the literature (Henderson 2004). The most interesting assumption in the BSOPM that has captured the attention of researches and practitioners alike is the constant volatility assumption. However, if the volatility is backed out from the BSOPM and plotted against different strike values, it displays a U-shape. Hence, the name 'volatility smile' (or skew). The graph deviated from a flat line, which contradicts the constant volatility assumption of the BSOPM.

Researchers then turned to more sophisticated methods for option valuation using stochastic volatility models (Ritchken and Trevor 1999; Peter and Kris 2004; Engle and Mustafa 1992; Duan 1995; Heston et al. 1997). The GARCH option pricing model (GOPM) introduced by Duan (1995) is based on a discrete-time model of the economy. The GOPM is derived based on the locally risk-neutralised probability measure where the option value can be calculated as a discounted expected value. In addition, the model allows the underlying asset return to follow a GARCH process (Bollerslev 1995), consequently eliminating the assumption of a constant volatility. This is indeed an attractive feature of the GOPM; however, the main drawback of the model is that it does not have a closed form solution. Therefore, Monte Carlo simulation is used to optimise the models parameters. Duan and Simonato (1998) have introduced an empirical martingale simulation that speeds up the Monte Carlo simulation process.

Researchers have also adopted different methods for pricing options, such as neural networks (Karaali et al. 1997). The vast majority of neural networks research has been focused on forecasting financial time series (Schittenkopf et al. 2000; Kaastra and Boyd 1996). Until the late 1990s, research on option pricing with neural networks was limited. To date, most research compares the performance of neural networks with the BSOPM option-pricing model (Meissner and Kawano 2001; Amilon 2003; Yao et al. 2000; Tino et al. 2001; Bennell and Sutcliffe 2004). Most research demonstrates comparable or slightly better performance of neural networks compared with the traditional option pricing models such as BSOPM. However, there has been no real attempt to compare the neural networks to more advanced models such as the GOPM. In Hanke (1997), a neural network was constructed to give an approximate price to GOPM. This was achieved by training the network on different input combinations for the GOPM allowing the neural networks to approximate the GOPM, thereby overcoming the numerical simulation issues of the GOPM.

In this chapter, we study the out-of-sample pricing and delta hedging capabilities of the GOPM, the BSOPM and two types of option pricing neural network designs. The first design is similar to the common approach where a neural network is trained on the option price. The second method is our proposed method which involves training the neural network directly on the BSOPM implied volatility and then using the BSOPM formula to derive the theoretical price.

We briefly review the option pricing models to increase the readability of this chapter.

7.1.1 GARCH Option Pricing Model (GOPM)

The GARCH Option Pricing Model (GOPM) introduced by Duan (1995) is based on a discrete-time model of the economy. The value of the index at time t, can be assumed to have the following dynamics:

$$r_t = r_f + \lambda \sigma_t - \frac{1}{2}\sigma_t + \sigma_t \varepsilon_t \qquad (7.1)$$

$$\varepsilon_t | \Omega_{t-1} \sim N(0,1)$$

$$\sigma_t^2 = \beta_0 + \beta_1(\varepsilon_{t-1} - \gamma)^2 \sigma_{t-1}^2 + \beta_2 \sigma_{t-1}^2 \qquad (7.2)$$

where λ is interpreted as the risk premium. To ensure that the variance is positive and stationary, the following constraints are applied:

$$\beta_0 > 0$$
$$\beta_1 \geq 0$$
$$B_2 \geq 0$$
$$\beta_1(1 + \lambda^2) + \beta_2 < 1$$

The unconditional variance is given by $\beta_0/(1 - \beta_1(1 + \lambda^2) - \beta_2)$. This process is reduced to the BSOPM homoscedastic lognormal process when $\beta_1 = 0$ and $\beta_2 = 0$.

It has been demonstrated by Duan (1995) that under the Local Risk Neutral Valuation Relationship (LRNVR) the conditional variance does not change. However, under measure Q, the conditional expectation of r_t is the risk free rate r_f.

$$E^Q[\exp(r_t)|\Omega_{t-1}] = \exp(r_f) \qquad (7.3)$$

To derive the GOPM, the risk neutral valuation relationship has to be generalised to the LRNVR:

$$r_t = r_f - \frac{1}{2}\sigma_t^{*2} + \sigma_t^* \varepsilon_t^*,$$
$$\varepsilon_t^* \sim N(0,1), \qquad (7.4)$$

$$\sigma_t^{*2} = \beta_0 + \beta_1(\varepsilon_{t-1}^* - \tilde{\gamma})^2 \sigma_{t-1}^{*2} + \beta_2 \sigma_{t-1}^{*2} \tag{7.5}$$

By having $\tilde{\gamma} = \lambda + \gamma$, the risk-neutral pricing measure is determined by four parameters: β_0, β_1, β_2 and $\tilde{\gamma}$. Using the above formulation, the asset terminal is then calculated at time T

$$S_T = S_t \exp\left(r_f(T - t) - \frac{1}{2}\sum_{i=t+1}^{T} \sigma_i^{*2} + \sum_{i=t+1}^{T} \sigma_i^* \varepsilon_i^* \right) \tag{7.6}$$

The terminal asset price is then calculated using Monte Carlo simulation. A set of N random path of residuals $(\varepsilon_{t+1,j}^*, \ldots, \varepsilon_{T,j}^*)$ is generated with J = 1 to N. The residuals are used to calculate the asset prices $S_{T,j}$. Using the terminal asset price series, the option price is then obtained by risk-neutral conditional expectation E^*:

$$C_{GARCH} = \exp(-r_f^*(T - t))E^*[\max(S_T - K, 0)] \tag{7.7}$$

The final option price is then approximated as follows:

$$C_{GARCH} = \exp(-r_f^*(T - t))\frac{1}{N}\sum_{j=1}^{M} \max(S_{T,j} - K, 0) \tag{7.8}$$

Rather than using standard Monte Carlo simulation, Empirical Martingale Simulation (EMS) is adopted. The EMS method has been shown to accelerate the convergence of the Monte Carlo prices estimates as demonstrated by Duan and Simonato (1998).

The Monte Carlo simulation is given by:

$$S_t(i) = S_0 \exp[r_f t]Z_t(i) \tag{7.9}$$

$$Z_t(i) = Z_{t-1}(I) \exp[-0.5\sigma_t^2 + \sigma_t \varepsilon_t(i)] \tag{7.10}$$

and for the EMS, $Z_t(i) = \dfrac{Z_{t-1}(i)\exp[-0.5\sigma_t^2 + \sigma_t\varepsilon_t(i)]}{\frac{1}{n}\sum_{i=1}^{n} Z_{t-1}(i)\exp[-0.5\sigma_t^2 + \sigma_t\varepsilon_t(i)]}$

The delta that corresponds to the GOPM is given by Duan (1995):

$$\Delta_t^G = \exp\{-(T - t)r\}E\left[\frac{S_T}{S_t}I(S_T, K)|\phi_t\right] \tag{7.11}$$

where I(S_t, X) = 1 is $S_t \geq$ K and 0 if S < K. Since there is no analytical solution for Δ_t^G the deltas are computed via Monte Carlo simulations.

7.1.2 BSOPM Option Pricing Model

The BSOPM theorem was first published in 1971 (Black and Scholes 1973). It is the most widely used pricing model. The model states that if S(t) is the asset price that follows a generalised Wiener Process

$$dS(t) = \mu S(t)dt + \sigma S(t)dz(t) \tag{7.12}$$

$z(t)$ is a Brownian motion, where the interest rate and volatility are constant. Then a call option on the asset, expiring at time T and with strike price K will have value at time t:

$$C(t, T) = S(t)N(d_1) - Ke^{-r(T-t)}N(d_2) \tag{7.13}$$

where, $d_1 = \dfrac{\ln\left(\frac{S(t)}{K}\right) + \left(r + \frac{\sigma^2}{2}\right)(T-t)}{\sigma\sqrt{T-t}}$ and $d_2 = \dfrac{\ln\left(\frac{S(t)}{K}\right) + \left(r - \frac{\sigma^2}{2}\right)(T-t)}{\sigma\sqrt{T-t}}$

$$N(x) = (2\pi)^{-1/2} \int\limits_{-\infty}^{x} \exp\left(-\frac{z^2}{2}\right) dz \tag{7.14}$$

$N(x)$ is the cumulative probability distribution for a standard normally distributed variable. The BSOPM delta is given by $N(d1)$, which also measures the sensitivity to the underlying asset.

7.1.3 Implied Volatility

The implied volatility is a volatility parameter, which equates the market price with the price given by the BSOPM formula. The implied volatility, $\sigma_t^{BS}(K, T)$ is a function of K (Strike) and T (time to maturity) (Hull 2003). The two most interesting features of the volatility surface, which have been studied and analysed by researchers, are the volatility smile (skew) and the term structure and the level of implied volatility changes with time. The volatility smile is a key indicator of an unrealistic assumption of constant volatility. Whereas, the changes in the implied volatility level with time is seen by the deformation of the volatility surface with time. Therefore, the ability to capture the deformation of the volatility surface will lead to accurate option pricing.

Cont and da Fonseca (2001) have expressed mathematically the implied volatility surface as follows:

$$C_{BS}\left(S_t, K, \tau, \sigma_t^{BS}(K, T)\right) = C_t^*(K, T) \tag{7.15}$$

where $\sigma_t^{BS}(K, T) > 0$.

The value of the call option as a function of the volatility is a monotonic mapping from $[0, +\infty$ [to] $0, S_t - Ke^{-rt}]$. The implied volatility $\sigma_t^{BS}(K,T)$ of a call option with strike K and price maturity of T is dependent on K and T. If K and T are fixed, $\sigma_t^{BS}(K,T)$ can be generalised to follow a stochastic process. For a fixed t, the value will depend on the options characteristics such as maturity and strike level K. Equation 6.17 represents the volatility surface at time t.

$$\sigma_t^{BS} : (K,T) \rightarrow \sigma_t^{BS}(K,T) \tag{7.16}$$

$$I_t(m,\tau) = \sigma_t^{BS}(mS(t), t+\tau) \tag{7.17}$$

where m is the moneyness, $I_t(m,\tau)$ is the implied volatility function.

The two most important features of this surface are volatility term structure (or volatility smile) and the changes in the volatility levels with time. Thus, the evolution in time of this surface will reflect the evolution of market option prices. Many previous studies have attempted to explain the information contained in the implied volatility (Day and Lewis 1992; Canina and Figlewski 1993; Christensen and Prabhala 1998; Blair et al. 2001; Jiang and Tian 2005) where the majority of studies have shown that the implied volatilities contain relevant measurement errors with quantifiable magnitude. Earlier research shows the implied volatility to be an inefficient forecast of volatility (Canina and Figlewski 1993). However, more recent studies such as (Jiang and Tian 2005) show that the implied volatility subsumes all information contained in the historical volatility and is capable of providing an accurate forecast of future volatility. This information can be utilised in many aspects since the option prices reflect the participant's expectations of the market's future movements. So, if the option market is efficient such that all information is observed and the correct option pricing model is specified, the implied volatility should subsume the information contained in other variables in explaining future volatility. That is, the implied volatility should be an efficient forecast of future trends over the life of the option.

7.1.4   Artificial Neural Network (ANN)

The ANN consists mainly of an input layer, one more hidden layer and an output layer. The layers are connected via a set of weights. The hidden layer and the output layer consist of individual neurons. The inputs are multiplied by the weights and a bias term is added which then constitutes the input to the activation function. This then serves as the input to the following layer. The activation of the output layer is given by:

$$F(x) = G\left(\sum_J w_{ij} H\left(\sum_K w_{jk} x_k + B_k\right) + B_J\right) \qquad (7.18)$$

The activation functions of the neurons could be chosen to be linear or non-linear functions. A sum of error-squared function is normally used as the objective function for the training of the MLP. So the MLP is trained to minimise this function with respect to the in-sample data.

$$E = \frac{1}{N} \sum_1^N (t_i - F_i(x))^2 \qquad (7.19)$$

The MLP performance is dependent on the initial values of the weights. To overcome this issue, the neural network is trained 50 times using different initial values for the weights. The weight set that introduces the least error is then adopted (Fig. 7.1).

7.1.5   Neural Networks for Option Pricing

This approach has been widely studied in the literature. This approach was also followed by Bennell and Sutcliffe (2004). The input data was partitioned according to the moneyness of the options in-the money (ITM), at-the-money (ATM) and out-of-the-money (OTM). For each set, a different network was used. This method has been proven to improve the pricing capabilities of the ANN (Yao et al. 2000). The in-sample and validation set for ATM and ITM were 168 days and 84 respectively. For OTM 64 and 20 days for the in-sample and validation set were used.

Fig. 7.1 Single layer artificial neural network

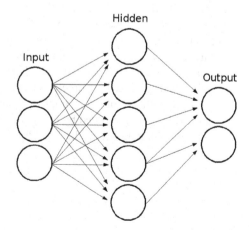

The homogeneity hint (Merton 1973) is used to simplify the input parameters (Garcia and Cirano 1998). The strike and index can be combined into a single input index/strike that is also interpreted as the measure of the option's moneyness. The other inputs used are the option maturity and historical volatility. The target set is the close price/strike. From this point onwards, this model will be referred to as NNp.

7.1.6   My Option Pricing Model

Given the limited success in applying neural networks to option pricing, we adopt a new method for training the neural network with the intention of capturing the underlying asset dynamics of the instrument which then can be translated to an option price using known market variables. This new methodology allows the neural network to capture the dynamics of the underlying asset and the changes in the volatility levels through time. Using the option price to train the neural network has been proven to be ineffective; therefore, instead of training the neural network to price the option directly, we allow the neural network to learn the implied volatility. So the neural network will be optimised on the implied volatility, thereby allowing the neural network to forecast the implied volatility for different maturities and moneyness. Many studies have demonstrated that vital information is contained in the implied volatility especially since the implied volatility contains the market expectation of the underlying asset future movements. To allow the neural network to capture the entire volatility surface, the data set was not partitioned by moneyness, thereby allowing the neural network full visibility of the volatility surface. This approach is a mapping exercise described in Eq. 7.18.

After experimentation with the different inputs, the following combination was selected for optimal performance. The in sample and validation set were 20 and 5 days respectively, and the output sample was 1 day. The inputs used were moneyness (index/strike), time to expiry and historical volatility. The target data set was the BSOPM implied volatility. Once the training of the ANN is completed, the implied volatility produced by the ANN would be plugged into the Black-Schole formula to obtain an analytical price of the option. This model would be referred to as NNiv. To further verify the adequacy and completeness of the proposed model, a delta hedging scenario is constructed. This will allow the model to be tested to ascertain whether it is able to predict the changes in the underlying asset, which will also verify the choice of using implied volatility for training the neural network. The delta values are generated by using the implied volatility forecast of the neural network and plugging it into the BSOPM delta formula.

7.2 Issues Investigated

Earlier we highlighted the strengths and weaknesses of the neural networks methods used in pricing options. In this section, we elaborate on the current issues that are experienced with pricing option with neural networks. In theory, neural networks should be able to outperform the any time series model as it is able to cater for non-linear behaviours in the underlying data. Also, it can approximate any continuous function up to a certain degree of accuracy. However, results in the literature show very weak performance of the neural networks in pricing options. For this reason, no study has been conducted so far to evaluate the option pricing performance relative to such models as the GOPM. Searching for a solution to this problem, we opted to model the key variable that can potentially explain the options' future behaviour. This is achieved by allowing the neural network to learn the implied volatility surface. The neural network will then be able to predict the future implied volatility levels. The implied volatility value produced by the neural network can then be converted to an option price using the BSOPM. We compare our proposed method to the pricing performance of other neural network methods that are based on the popular approaches documented in the literature. To provide a better understanding of the neural networks' capabilities, we benchmark these methods with BSOPM and GOPM. This will be the first study to evaluate the pricing capabilities of the GOPM to neural networks. This is necessary to evaluate the neural networks' accuracy compared to sophisticated pricing models. This study is then extended to evaluate the models in delta hedging scenarios. This will verify the completeness of the model and whether it is able to predict changes in the underlying asset. In this study, we:

- Address the weak option pricing capabilities of the neural networks as observed in the literature
- Derive a new method for pricing option pricing via neural networks
- Demonstrate superior option pricing capabilities of neural networks relative GOPM and BSOPM
- Evaluate the hedging performance of the neural networks to BSOPM and neural networks.

7.3 New Option Pricing Model and Solution Overview

The BSOPM was the first model to analytically calculate a fair option price. This model enjoyed wide popularity and was adopted by practitioners due to its simplicity and its ability to produce an option value instantaneously. The model derivation was based on many unrealistic assumptions. It was soon realised that these assumptions had adverse effects on the model's pricing capabilities which is apparent in the model's pricing biases. Many variations of this model emerged as improvements on the BSOPM by addressing some of the model's assumptions. The GOPM has gained popularity as it allows the volatility process to follow a GARCH process, thereby overcoming the constant volatility assumption. The drawback to the GOPM is the absence of a closed form solution, which means the GOPM parameters are optimised using option data using the Monte Carlo simulation. This is a computationally intensive process which also limits it application in real-life trades.

The constant volatility assumption of the BSOPM has captured the attention of researchers and practitioners alike. This is due to the nature of volatility not being directly observable in the market and is known to change over time. Also, when the implied volatility is plotted against strike prices, it takes on a U-shape. This gives a forward looking view derived from the option prices, which indicates how the market is anticipating the volatility movements. This behaviour shows valuable information that can be derived from the implied volatility. Modelling option behaviour over time will be dependent on the model's ability to capture the implied volatility process.

Neural networks have been applied to the option pricing problem with the aim of overcoming some of the pricing biases of the BSOPM. The common approach adopted in research is to feed the neural network with similar inputs to the BSOPM, and use the market option price as the target value for training the neural network. Different variations of this method have been adopted to improve the pricing accuracy of the neural networks. However, all research thus far displays pricing improvements on the BSOPM. It is very interesting to see that no major work has been done to compare the pricing performance of the neural network to more advanced models such as the GOPM. The reason for this is the pricing improvements of the neural networks over the BSOPM are less significant for the GOPM. This means that the neural networks' performance does not stack up against the GOPM. For this reason, to date no work has been done to benchmark the neural network option pricing capabilities against more sophisticated option pricing models such as the GOPM. Neural networks are capable of superior option pricing performance if the correct neural network model is applied. This serves as our motivation to investigate the shortcomings of models presented in the literature and provide a neural network model to overcome these weaknesses.

The option pricing methods adopted for neural networks thus far has been to map some predefined inputs to the option price. The idea behind this method is to allow the neural network to mimic the BSOPM. Many variations of this method have been developed to improve the pricing accuracy. For instance, substantial pricing performance gains are achieved when the training data set is partitioned by moneyness and different neural networks are used for each of the training sets. However, the increase in pricing accuracy is still not sufficient to match that of the GOPM. This method suffers from major drawbacks that can impact on the performance of the neural network. Training the neural network on the option price is a cumbersome task. The option itself can suffer from many issues such as stale prices due to infrequent trading (Rubinstein 1985). Therefore, using the option price to guide the training of the neural network can be misleading as the neural network will attempt to map the inputs to an output that has poor correlation with the input. The BSOPM generates an option price using the inputs and there is no dependence on the option price. The neural network pricing accuracy in this case is very much dependent on the option price. In addition, the option price does not seem to be able to explain the underlying dynamics of the generating process. To overcome these issues, we deviate from the literature and search for a more effective approach to training the neural networks.

Neural networks are known for their generalisation and approximation properties. So rather than learning the option price, we allow the neural network to learn the behaviour of the option price through time. The main variable that can explain this key behaviour is the implied volatility, since the implied volatility is a forward looking tool for market anticipation. By allowing the neural network to capture the implied volatility surface through time, the neural network will be able to forecast future movements of the implied volatility across different strikes which can then be converted to an option price. To demonstrate the effectiveness of this method, we evaluate four models: BSOPM, GOPM and two neural network models. The first neural network model is in line with the methods used in the literature. The data set is partitioned by moneyness and different neural networks are trained on each set. The second neural network model consists of our proposed approach. Each model will be trained on the FTSE 100 European options and the performance of each model will be evaluated according to the out of sample pricing performance. With this experiment, we make the following contribution:

- Introduce a new neural network option pricing method via learning implied volatility
- Demonstrate the effectiveness of our approach to that used in literature
- Demonstrate the option pricing improvements of our method relative to other neural network and other option pricing models
- Provide the first study to compare neural network option pricing with GOPM
- Explain the weak neural network option pricing results presented in the literature.

7.4 Data

The data used in this research consists of European call options on the FTSE 100 index traded at the London International Financial and Options Exchange (LIFFE). The data for this research was obtained from SIRCA (http://www.sirca.org.au/), which covered a two-year period starting from 2/1/2000 and ending on 31/12/2001. The sample was made up of 63,094 call options and the daily Index value (adjusted for dividends).

The following constraints were applied to filter the data series: moneyness (index/strike) outside [−1.01, 0.9], maturity greater 175 days and less than 7 days and close price less than five where removed. The data series was then reduced to 26533 options. GB Libor rates were used for the risk-free rate inputs to the models.

The parameters for the GOPM and BSOPM were estimated on a daily basis. However, when optimising the ANN, the data was split into 3 sets: in-sample, validation and out-of-sample. The in-sample set is used as inputs to the ANN, thus impacting on the weights. The validation set is used to evaluate the error at each epoch. The training of the network is terminated when the error of the validation increases. The out-of-sample is data set is used for evaluating the ANN model. The hedging analysis was done over a one-month horizon for the month of February-2000.

7.5   Experimental Design

The aim of this research is to investigate the capability of neural networks to capture the market volatility dynamics, and hence accurately price call options. The competing models are the GOPM, two Neural Networks trained on the option price (NNp), and implied volatility (NNiv) respectively. The models are then compared based on the out-of-sample pricing accuracy and the hedging performance.

7.5.1   GOPM Parameters

The GOPM parameters were estimated for each day of the data sample. This was achieved by minimising the average sum squared error (Aboura 2005) (Lehnert) over all options on day t, with parameters from day t−1 as the initial values (Lehar et al. 2002).

$$
\text{Min } SSE(\Theta) = \sum_{t=1}^{T} \sum_{n=1}^{N} \left(\frac{\left(\widehat{C_{t,n}} - C_{t,n} \right)}{C_{t,n}} \right) \tag{7.20}
$$

where $\Theta = \{\beta_0, \beta_1, \beta_2, \lambda\}, \widehat{C}$ and C are the model theoretical price and the actual call price respectively. The number of Monte Carlo simulation was initially set to 15,000. This was sufficient to generate stable estimates. In some cases, the number of simulations had to be increased to 30,000.

7.5.2 *Neural Network Training Methods*

In this research, two neural networks option pricing methodologies are examined; the first method is commonly used in research where the neural network is trained on the option price. This will allow the neural network to directly forecast an option price based on the inputs provided to the neural networks. In the second method, the neural network is trained using the BSOPM implied volatility as the target output. This method allows the neural network to predict the future implied volatility, which also means that the neural network can produce the implied volatility surface over any trading horizon. The implied volatility can be converted to an option price by plugging it into the BSOPM. In both methods, the neural networks were optimised by varying the number of inputs, length of in-sample and validation sets, and the number of hidden units. The combination that gives the lowest error is considered to be the optimal model and is chosen accordingly. The initial weight values were initialised randomly, and to eliminate the dependency on the initial weight values, this process was repeated 50 times.

7.6 Performance Measures

The first measure of the model performance is its ability to price options out-of-sample. However, this measure does not give a clear indication of the model's ability to generate profits. Therefore, the hedging performance of the model is a much better indicator of the model's ability to capture the underlying asset dynamics in the market.

7.6.1 *Pricing Accuracy*

To analyse the model's performance, Relative Pricing Error (PE) and Absolute Relative Pricing Error (APE) were calculated for each model.

$$RPE = \frac{1}{n} \sum_{i=1}^{n} \left(\frac{\widehat{C}_i - C_i}{C_i} \right) \tag{7.21}$$

$$ARPE = \frac{1}{n} \sum_{i=1}^{n} \left| \frac{\widehat{C}_i - C_i}{C_i} \right| \tag{7.22}$$

where $\widehat{C}$ and C is the model theoretical call price and the actual call price. The errors are reported against maturity and moneyness.

7.7 Delta Hedging

Most papers consider a hedge portfolio with one unit short in an option, Δ units of the underlying asset and the rest in a risk free asset (Schittenkopf and Dorffner 2001).

7.7.1 Solution Overview

If the option pricing model is correctly specified, it is capable of predicting changes in the underlying asset. This is achieved by taking the partial derivative of the option price relative to the underlying asset. Delta hedging can be applied to verify the effectiveness and completeness of the model. In this research, we will further evaluate the option pricing models by carrying out delta hedging exercises for each of the models. The delta hedging process applied here involves replication of a portfolio of one call option and holding the delta amount in cash to earn the risk free rate. If the option pricing model is providing the correct delta value, which means it is able to correctly predict the changes in the underlying asset, then the hedging error should be minimal. Ideally, hedging should be done as frequently as possible. However, in reality this is not feasible due to the transaction costs associated with the execution of a trade. A generally high period of volatility will require the portfolio to be replicated more frequently which can offset any profits gained.

In this research, we assume that there are no costs associated with the hedging and the portfolio is replicated as needed. The hedging exercise is conducted by pricing the option using the option pricing model. If the theoretical option price is greater than the market price, the option is bought and a delta amount in shares is sold. The money gained from the sale of the shares is placed in the bank to earn the risk free rate. If the theoretical price is less than the market price of the option then it is sold, and the delta value of the underlying asset is bought, and the remaining cash is put in the bank to earn the risk free rate. The portfolio is evaluated on a daily basis and readjusted accordingly until the expiry of the option. At expiry, we calculate the delta hedging error over the life of the option, which will be used to evaluate each of the option pricing models. By carrying out this exercise, we make the following contributions:

- Providing the first study to evaluate the hedging performance of neural networks against the performance of the GOPM.
- Evaluate the hedging performance of our method compared with the GOPM and BSOPM.

7.7.2 New Method Developed

In this research, a hedge portfolio is constructed in a similar fashion to that adopted by Amilon (2003) and Vahamaa (2004). That is, we buy options that are underpriced (where theoretical option price is higher than market price) and sell the options that are overpriced. For underpriced options, we construct the following portfolio:

We buy option $V_0^C = C_0$ and sell index $V_0^I = -I_0 \Delta_0$ and put the rest in risk-free asset, $V_0^B = I_0 \Delta_0 - V_0^C$. The portfolio is then replicated as follows:

$$V_t^I = -I_t \Delta_t$$

$$V_0^C = C_0$$

$$V_t^B = \exp(rt) V_{t-1}^B + I_t (\Delta_t - \Delta_{t-1})$$

The value of the portfolio at time T is given by

$$V_T = V_T^B + V_T^I + V_T^C \tag{7.23}$$

which gives us the absolute hedging error

$$\xi = \exp(rt)|V_T| \tag{7.24}$$

where, V_t^B, V_t^I and V_t^C are the amount invested in the risk free asset, index and call option at time t.

The delta hedging exercises are carried out for a period of one month (6/2/2000 to 6/3/2000).

7.8 Results

Below are the estimated mean values of the GOPM parameters (Table 7.1).

The risk premium is extracted from the option prices rather than from the index time series. It was found to be a little weak which in turn will have a minor impact on the option price. The annual volatility was found to be approx. 22.37% which is a close approximation to the historical volatility when estimated on the index prices 20.7%.

Table 7.1 GOPM average parameters

β_0	β_1	β_2	γ	$\sqrt{\sigma}$
3.109E−05	1.848E−01	3.184E−01	6.745E−01	22.37%

The performance measures used for all models are relative pricing error (RPE) and the absolute relative pricing error (ARPE). A positive or negative value of PE indicates overpricing or underpricing of the model respectively. Table 7.2 shows the RPE for all models. On average, the GOPM and the NNiv overprice call options, whereas the BSOPM and NNp underprice the options. The NNp tends to overprice ST options and underprice options in the LT and MT range. The GOPM and NNiv have similar behaviour since both models overprice options over all maturities. The BSOPM seems to underprice options over all maturities. In addition, it has the worst mispricing performance across all maturity and moneynees.

The GOPM and NNiv models overprice the LT options across all moneyness. The worst mis-pricing performance of the GOPM model has a RPE of 0.069 for LT-DITM contracts. This behaviour is also seen for NNiv, where the model has the worst mispricing for LT-DITM contracts with a RPE of 0.075. For BSOPM the worst performance for ST-OTM options with a RPE of –0.901 and the NNp has a PE of 0.837 for ST-DOTM contracts.

Table 7.2 Relative pricing errors for all models

Moneyness	Model	Time to maturity			
		LT	MT	ST	Total
DOTM	GOPM	0.021	−0.010	−0.020	−0.009
	NNp	−0.062	−0.047	0.837	0.128
	BSOPM	−0.098	−0.011	0.038	−0.005
	NNiv	0.034	0.114	−0.024	0.073
OTM	GOPM	0.027	−0.007	0.059	0.024
	NNp	−0.145	−0.365	0.269	−0.085
	BSOPM	−0.497	−0.691	−0.901	−0.762
	NNiv	0.005	0.014	0.024	0.017
ATM	GOPM	0.039	0.008	0.030	0.021
	NNp	−0.133	−0.083	0.238	0.044
	BSOPM	−0.272	−0.240	−0.215	−0.232
	NNiv	0.025	0.004	0.007	0.007
ITM	GOPM	0.059	0.013	−0.008	0.007
	NNp	0.063	−0.013	0.142	0.066
	BSOPM	−0.165	−0.084	0.002	−0.049
	NNiv	0.040	0.008	−0.019	−0.002
DITM	GOPM	0.069	0.059	0.044	0.057
	NNp	−0.088	−0.104	0.010	−0.051
	BSOPM	−0.098	−0.011	0.038	−0.005
	NNiv	0.075	−0.002	−0.020	0.004
All	GOPM	0.045	0.018	0.042	0.030
	NNp	−0.109	−0.199	0.350	0.004
	BSOPM	−0.584	−0.900	−0.641	−0.767
	NNiv	0.022	0.043	0.009	0.027

The APE figures are displayed in Table 7.3. On average, the performances of GOPM and NNiv models increase with maturity. Again, the BSOPM performs the worst on all accounts. On average, the NNiv has a slightly better performance than does the GOPM. The GOPM model performs slightly better for MT contracts and the NNiv has a better performance for ST and LT contracts. All models have performed the worst for ST-DOTM contracts. In addition, the GOPM performs the best for LT-DITM, whereas the NNiv and BSOPM are at their best for ST-DITM contracts. The NNiv seems to perform best for the MT-ITM contracts. NNp will be excluded from the hedging analysis due to the unsatisfactory performance with the option pricing. The rest of the models are compared over a 1-month horizon, where delta hedging was performed until the option expiry. Then absolute hedging error Eq. 6.24 was used as the benchmark statistic for all models. As shown in Table 7.4, the BSOPM and NNiv on average are very close with a difference of 0.03. But the NNiv on average performed better for the ITM options. The GOPM has the worst performance on all accounts. For ATM-ST and ITM-ST options, the NNiv

Table 7.3 Absolute relative pricing errors for all models

Moneyness	Model	Time to maturity			
		LT	MT	ST	Total
DOTM	GOPM	0.179	0.246	0.384	0.267
	NNp	0.590	0.513	1.014	0.625
	BSOPM	0.993	1.815	1.289	1.584
	NNiv	0.167	0.269	0.358	0.273
OTM	GOPM	0.105	0.157	0.272	0.202
	NNp	0.519	0.483	0.711	0.580
	BSOPM	0.502	0.713	0.984	0.809
	NNiv	0.102	0.154	0.262	0.195
ATM	GOPM	0.078	0.100	0.139	0.116
	NNp	0.263	0.212	0.400	0.295
	BSOPM	0.274	0.262	0.277	0.270
	NNiv	0.086	0.094	0.135	0.112
ITM	GOPM	0.082	0.074	0.079	0.077
	NNp	0.154	0.131	0.232	0.181
	BSOPM	0.169	0.120	0.077	0.104
	NNiv	0.081	0.071	0.075	0.074
DITM	GOPM	0.039	0.061	0.050	0.053
	NNp	0.160	0.143	0.151	0.149
	BSOPM	0.099	0.074	0.051	0.069
	NNiv	0.098	0.066	0.058	0.068
All	GOPM	*0.121*	*0.167*	*0.233*	*0.188*
	NNp	*0.491*	*0.440*	*0.674*	*0.528*
	BSOPM	*0.589*	*0.924*	*0.742*	*0.818*
	NNiv	*0.119*	*0.172*	*0.222*	*0.185*

Table 7.4 Discounted absolute hedging error

	Model	LT	MT	ST	Total
ATM	NNiv	0.38	19.56	**15.66**	15.20
	BSOPM	**0.29**	**19.09**	15.74	**15.06**
	GOPM	0.41	22.38	16.33	16.56
ITM	NNiv	0.41	23.37	**27.47**	**22.83**
	BSOPM	**0.31**	**21.99**	29.65	23.22
	GOPM	0.42	25.93	30.62	25.39
OTM	NNiv	0.23	9.91	6.70	7.57
	BSOPM	**0.20**	**9.90**	**5.89**	**7.28**
	GOPM	0.30	12.20	6.52	8.72
Total Nniv		0.33	16.67	17.03	14.90
Total BSOPM		**0.26**	**16.11**	**17.55**	**14.87**
Total GOPM		0.37	19.18	18.27	16.56

performed better to the reset of the models. In addition, NNiv seemed to get very close to the BSOPM. For instance, the OTM-MT option, the BSOPM and NNiv were in fact very close with a difference of 0.01. However, the GOPM displayed far worse performance with big differences when compared to the BSOPM. The NNiv was clearly outperformed by the GOPM on all accounts expected for OTM-ST but with a difference of 0.18.

The numbers shown in bold in the Table 7.4 indicate the most performant model with the smallest hedging error. The models were also tested with respect to their hedging performance. The GOPM performed the worst. However, the BSOPM on average outperformed the others except for the ITM options where NNiv did better. The NNiv did outperform the BSOPM in some categories and in most cases was trailing very closely. However, it did outperform the GOPM. There are a couple of factors that may have affected the outcome of the results. Firstly, the exclusion of options based on moneyness and maturity could have resulted in the loss of significant information. Also, the hedging strategy of keeping the option till maturity could skew the results. The method described in (Amilon 2003) would have been a better option. In addition, the use of the discounted absolute hedging error alone does not provide enough statistics to evaluate the model performance.

To further analyse the models pricing errors, the following regression is performed.

$$ARPE = aTTE + bMon + c + \varepsilon \qquad (7.25)$$

$$\varepsilon \sim N(0, \sigma^2)$$

The ARPE is regressed on TTE (Time to Expiry) and Mon (Moneyness). The estimated coefficients are displayed in Table 7.5. The coefficients R^2 and F-test for NNiv and the GOPM are similar. This indicates that the models price the options in a similar manner. For all models, the ARPE is smaller for longer maturities and

Table 7.5 Regression results

	NNiv	GOPM	NNp	BSOPM
C	2.097	2.052	4.324	13.931
	(−46.116)	(−45.072)	(−16.508)	(69.214)
A	−0.327	−0.330	−0.645	−0.769
	(−26.065)	(−26.306)	(−9.960)	(−13.869)
B	−1.900	−1.851	−3.797	−13.372
	(−40.983)	(−39.865)	(−14.103)	(−65.148)
R^2	14.66%	13.71%	2.65%	25.05%
F-test	1047.439	1014.297	130.631	2133.957

increased moneyness. Therefore, ITM options are priced more accurately than the OTM options and the models price LT contracts with better accuracy than ST contracts.

7.9 The Empirical Dynamics of the Volatility Smile

In this section, we study the behaviour of the implied volatility of the GOPM and the NNiv over time. The implied volatility of NNiv is given directly as the output of the model, whereas the GOPM produces a theoretical option price. To extract the implied volatility from the GOPM, the option prices are equated to the BSOPM formula and the implied volatility is backed out.

To study the volatility smile, a theoretical series is generated by an out-of-sample fit of the GOPM parameters calculated on the previous day. The implied volatility is then backed out for each option price using the BSOPM formula. Figure 7.2 shows the first day out-of-sample forecast. It can be seen that the volatilities are higher for the shorter maturities. As the option moves from ITM to ATM, they converge and the slopes are reversed. This has been highlighted by Aboura (2005), where this point is of importance because the model is able to classify the volatilities according to it moneyness.

Figure 7.3 displays the implied volatility for the end of the month using the same parameters as those for the first day of the month. The deformation of the smiles is obvious. The short maturity has a higher implied volatility; however, the skew of the longer maturities becomes more linear.

The theoretical volatilities calculated for NNiv at the first day of the month are displayed in Fig. 7.4. The volatility smile is apparent in the graph. The graph also shows a similar behaviour to the skews generated by the GOPM. For all maturities, the slope of the skews changes as it moves from ITM to OTM. So, as the option approaches ATM, the volatilities across all maturities become similar. For the ITM-LT option, the skew tends to be linear; then as it moves past the ATM point, the NN gives higher volatilities. For ITM-ST options, the NN generates much larger volatilities than the ITM-LT options. As ST option passes the ATM range, the slope

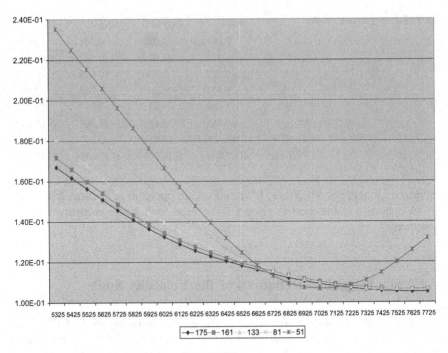

Fig. 7.2 GOPM out-of-sample skews for 4/9/2000

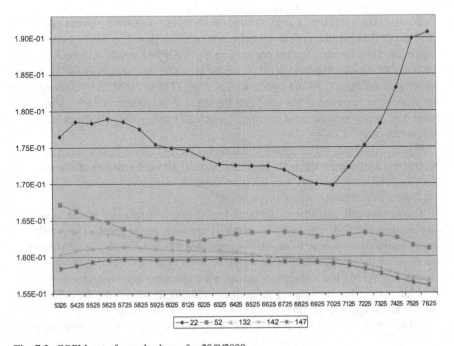

Fig. 7.3 GOPM out-of-sample skews for 28/9/2000

changes direction forming the smile. However, the volatilities are smaller than the LT for OTM options.

In Fig. 7.5, there does not seem to be any deformation of the skews. The skews also follow the same behaviour as those in Fig. 7.4. The advantage of the GOPM model is its ability to capture the deformation of the skew through time. This shows that the model is sensitive to changes in the underlying asset. The GOP model parameters are calibrated under LRNVR conditions by simulating the underlying asset process and pricing the options at expiry accordingly. However, the NN is designed and trained differently. The ability of the NN to price options out-of-sample is a key indicator of the ability of the NN to capture the dynamics of the FTSE options. The NN trained on mapping the input space (TTE, Moneyness, Historical Volatility) to Black-Scholes implied volatility. Due to the high number of variables in the NN, 1 day of historical data was not sufficient to produce a stable result. Hence, the NN was trained on 20 days of data, which contains substantial information. Therefore, when the smiles are generated out-of-sample towards the expiry of the options, the neural network is able to display the actual skew at that point in time. This is a good indication that the NN is a suitable tool for long-term forecasting. Also, its ability to generate the skews out-of-sample could be an advantage when performing delta hedging.

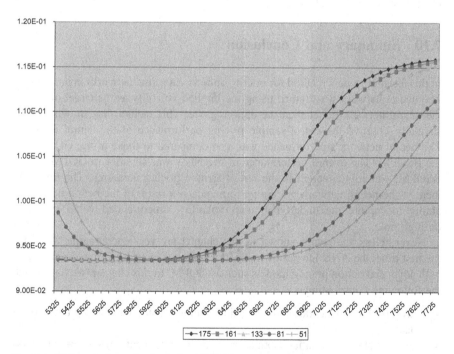

Fig. 7.4 NNiv out-of-sample skews for 4/9/2000

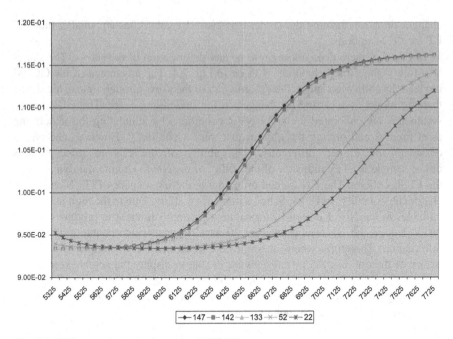

Fig. 7.5 NNiv out of sample skews for 28/9/2000

7.10   Summary and Conclusion

In this research, a new method for pricing options via neural networks is introduced.
The neural network was trained using the implied volatility as the target output of
the neural network rather than the options price. This method has shown to substantially improve the out-of-sample pricing performance of the neural network.
The neural network's performance was also compared to those of the GOPM and
the BSOPM. The overall performance of the NNiv was superior to GOPM and to
BSOPM, which is apparent in the out-of-sample pricing measures. The improvement in the neural network pricing performance is a result of the neural network's
ability to capture the underlying asset volatility structure and volatility levels
through time.

Nevertheless, the results shown in this chapter are very encouraging, since for
the first time, the ANN have shown superior option pricing performance compared
with advanced option pricing models such as GOPM. In this chapter, we have made
the following contributions:

- Introduced a new neural network option pricing method by learning the implied
  volatility.
- Demonstrated substantial option pricing improvements using our method
  compared with methods presented in the literature.
- Provide the first study to compare neural network option pricing to GOPM.

- Provide the first study to evaluate hedging performance neural networks to GOPM.
- Evaluated the hedging performance of our method against the GOPM and BSOPM.
- Provide the first study to demonstrate the superior accuracy of neural networks to GOPM.

By demonstrating the neural network's superiority over the GOPM, and its ability to produce an option price instantaneously, we show that it can be deployed as a real-time trading tool. An ongoing direction of this research is to extend the current findings to train the neural network on the BSOPM delta rather than using the neural network's implied volatility, and to examine different types of neural networks such as RBF which may produce improved results.

Chapter 8
Value-at-Risk

The increase in volatilities in world markets has made risk management tools a necessity for understanding and controlling risk exposure. VaR models have been designed and adopted by finance institutions as the standard tool for reporting risk exposure. The popularity of a VaR model stems from its ability to summarise and present the worst expected loss of a portfolio over a time period as a single number. This measure can be easily understood by senior management without necessarily understanding the mathematical reasoning behind it. The wide adoption of VaR models by financial institutions is also attributed to J.P. Morgan's development of VaR methodologies. In addition, the Basle Committee on banking supervision at the Bank for International Settlements has made it a requirement for banks to establish minimum capital requirements based on the VaR measures subject to certain criteria. There are different methods for calculating the VaR (see appendix B), no matter which approach is considered, accurate measurement of the exposure to market risk is vital to the effective application of the VaR models in risk management.

8.1 Value-at-Risk Review

Value-at-Risk (VaR) concepts can be traced back to the late 1980s where financial firms began to adopt VaR to measure the risk exposure of their trading portfolios. J.P. Morgan released the *RiskMetrics^{TM}* in October 1994 in an attempt to standardise the application of VaR in industry. This significantly contributed to the growth in the popularity of VaR as a risk assessment tool. VaR has also attracted major attention in academia, where its application has been extended into many other risk areas. The ability to summarise the risk exposure as a single number has allowed VaR to become applicable across different levels, where managers do not

© Springer International Publishing AG 2017
F. Mostafa et al., *Computational Intelligence Applications to Option Pricing, Volatility Forecasting and Value at Risk*, Studies in Computational Intelligence 697, DOI 10.1007/978-3-319-51668-4_8

need an in-depth statistical understanding of the risk models to understand the VaR measure, since it provides a single number that expresses the risk exposure in dollar terms for a given time horizon and confidence level, that can be easily interpreted and understood.

Given the recent volatility in local and world markets, empirical modelling has been widely adapted in many areas of risk management. Despite the increase in their application and popularity, there is little evidence that these models actually work. If the risk models perform their job effectively, no systematic failures will be observed. The risk models suffer from limitations inherent in their design. These limitations become more apparent during a period of crisis when the basic statistical properties of the market differ substantially compared to stable periods. VaR models have become a popular risk measurement tool in finance as they attempt to capture the potential market risk of an assets portfolio. This is achieved by measuring the maximum loss of a portfolio value over a given time period over a time horizon. Despite the wide adoption of the VaR models by financial institutions, VaR has been criticised for its simplification of the risk measure and the unrealistic assumptions. The main concept of the VaR model is to produce a quantitative figure of a portfolio's downward risk. This provides a consistent measure of risk for different positions and instruments (Dowd 1999). It also accounts for correlations between different risk factors which is important when computing risk measures of a portfolio with more than one instrument.

The main criticism of the VaR model is that it reduces all information down to a single measure where potential vital market information could be lost in the process. Also, the VaR model does not give any further information on the extent of potential losses that can occur beyond the VaR estimate. Another drawback of the VaR model is that different portfolios' risk exposure cannot be compared using the VaR estimates; that is, two portfolios with the same VaR estimates do not carry the same risk (Tsay 2005). VaR models are typically based on assumptions which are not necessarily valid. For this reason, the VaR models are not complete models and can break down in extreme circumstances, especially when they are expected to work. Hence, VaR models should be used together with other risk management techniques to obtain a better understanding of the risk exposure (Tsay 2005).

There are three main methods of calculating the VaR: Variance-covariance, Historical and Monte Carlo simulation methods. Below is a summary of the main advantages and disadvantages of each method.

8.2   Value-at-Risk Definition

VaR was developed essentially to measure the market risk of a portfolio. Market risk is mainly caused by the volatility or the price movements of the asset price. So the VaR models in simple terms states the following:

Fig. 8.1 Illustration of VaR
confidence intervals

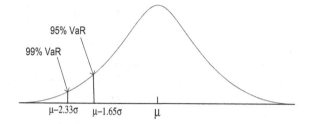

We are X percent certain that we will not lose more than V dollar in the next N days, (Hull 2003).

The concept of VaR gives a quantitative measure of the portfolio's downward risk. The important features of VaR include a common consistent risk measure for different instrument types in the portfolio. Also, it accommodates the correlation between different risk factors which is vital when computing risk measures for portfolios with more than one instrument. The calculation of VaR can be simplified if the distribution of the asset returns is assumed to be normal:

$$VaR_\alpha = \alpha \sigma W_0 \sqrt{N} \qquad (8.1)$$

α is the confidence interval, σ is the standard deviation of the returns and W_0 is the initial portfolio value and N is the time horizon measured in days as shown in Fig. 8.1.

It is essential to keep in mind the time horizon and the confidence level when analysing the VaR measures. It is also essential to take into account the method used for calculating the VaR such as the normal approximation of the asset returns. It is recommended to use the normal approximation for short time horizons such as daily VaR calculations. The confidence level must also be chosen to suit the problem at hand. For instance, the confidence level will depend on the risk aversion of the financial institution. If the institution is risk averse, then a high confidence level should be used.

8.3   Modelling Value at Risk with Neural Networks

There are two types of VaR calculations: parametric and non-parametric. The parametric models are based on statistical parameters of the return distribution. The non-parametric are based on simulations or the use of historical data (Ammann and Reich 2001). The three most popular methods for VaR calculations are discussed below.

8.3.1   *Historical Simulation Method*

The historical method is the simplest method for calculating VaR since it requires few assumptions about the distribution of the portfolio returns. This method makes use of historical data to construct the future profit and loss distribution of a portfolio which can be utilised to calculate the VaR measures. The distribution is constructed by subjecting the portfolio to daily changes of the past N days. This applies the past 100 days of market changes to the portfolio for the construction of N hypothetical mark to market profits and losses. Using these hypothetical profit and losses, VaR can then be determined.

The historical simulation method can be explained in the following steps:

- Identify the market factors relevant to the portfolio in question
- Obtain the historical values for the market factors for the past N days
- Subject the portfolio to the changes of past N days, by calculating the profit and loss of the portfolio using the changes in the market for the past N days
- Order the portfolio profit and losses in descending order
- Based on the desired confidence interval select the appropriate value, for instance, when N is 100 days, then the 95% VaR would be the 5th largest loss.

The use of actual historical prices eliminates the need to make assumptions about the underlying asset distribution. This also caters for fat tails which eliminate the normal assumption. The historical simulation method has many attractive features; for instance, it caters for non-linear pay-offs such as options. This method eliminates the need for complex covariance calculations.

The choice of the length of data set to be used in this method has to be considered carefully. The historical period should be long enough to produce a reliable estimate, but short enough to avoid "paradigm shift". However, if high confidence intervals are required, then a large data set is necessary. In this case, the data set might contain excess or extreme market movements that do not reflect the current market conditions (Manfredo and Leuthold 2001). Historical simulation ignores the time variance and assigns equal weights to events (Linsmeier and Pearson 2000). This means that this method does not react fast enough to the current market conditions. Also, it does not cater for high and low periods of volatility.

Historical simulation is the simplest of the VaR methods to implement. This method does not make any assumptions about the statistical distribution, volatility or correlations. The VaR value is derived directly from the returns series and can be applied to any instrument (Jorion 2001). However, this method has many shortcomings. For instance, it will not be useful for new instruments where there are insufficient data. Since the VaR value is calculated based solely on historical returns, the VaR estimate will be unreliable. The fundamental assumption underlying asset the historical simulation method is that the historical asset return movements remain unchanged in the future. So a VaR forecast can be calculated based on the historical return. In some cases, this can lead to distorted VaR estimates (Dowd 1999) due to the possibility that some occurrences are not captured in the historical data set such as high volatility periods which can lead to major losses. There is a trade-off associated with this method that is associated with the length of the historical data series. It is important to have a large time series to ensure stable VaR estimates; however, this method does not give any weighting or priority to recent observations (Jorion 2001). All observations have equal weighting where the VaR estimates will not react quickly to recent market data especially in cases of large market jumps. This method does not represent the return dynamics which can cause major potential losses when applied in a risk management scenario.

8.3.2 Variance-Covariance Method

The variance-covariance method is based on the assumption that the asset prices are normally distributed. The assumption of normality simplifies the implementation of this method significantly. After estimating the parameters of the return distribution for the entire portfolio, the VaR measure can be found by determining the loss value that matches or exceeds the specified percentage for a given time frame. Using the variance-covariance method, the VaR model can be expressed in terms of the variance-covariance matrix for the asset returns, where the sensitivity of the portfolio to price shifts can be calculated. The variance-covariance approach can be constructed using the following steps:

- The variance-covariance matrix is calculated using the historical returns. Let us consider a portfolio with two assets A and B. The variance-covariance is given by:

$$M = \begin{bmatrix} \sigma_A^2 & \sigma_{A.B} \\ \sigma_{A.B} & \sigma_B^2 \end{bmatrix} \tag{8.2}$$

$\sigma_{A.B}$ is the covariance between series A and B, σ_A^2 is variance of series A and σ_B^2 is the variance of series B.

- Calculate the sensitivities of the portfolio. This is the amount that a portfolio value will change for a predefined change (i.e. 1% change) in the underlying asset prices. This is achieved by examining the movements in each of the market prices separately.
- Calculate the volatility of the total change in the portfolio value.

$$v = \sqrt{\delta' M \delta} \text{ or} \tag{8.3}$$

$$v = \sqrt{\begin{bmatrix} \delta_A & \delta_B \end{bmatrix} \begin{bmatrix} \sigma_A^2 & \sigma_{A.B} \\ \sigma_{A.B} & \sigma_B^2 \end{bmatrix} \begin{bmatrix} \delta_A \\ \delta_B \end{bmatrix}} \tag{8.4}$$

where, δ a vector of market prices sensitivities, M is the variance-covariance matrix of the market prices.

- To calculate the VaR measure, the volatility v must be multiplied by the scaling factor (based on the desired confidence level). For instance, if we desire a 95% confidence level, then the VaR can be calculated as follows:

$$VaR_{95\%} = 1.645v\sqrt{T} \tag{8.5}$$

T is the holding period in days.

The assumption of normality is one of the main drawbacks of this method. This method has a tendency to produce fat tails in the distribution of the returns. This is a concern since fat tails underestimate the true risk associated with the portfolio. Also, a portfolio with option-like assets can produce inaccurate results due to the non-linear pay-offs. A portfolio with moderate option-like content can make use of the variance-covariance method as long as the holding period is very short (Manfredo and Leuthold 2001). This method is easily implemented and it is not computationally intensive. Also, the correlations and standard deviations can be derived to determine the impact on the risk variables.

This approach is based on the assumption of changes in the market and portfolio values are normally distributed. This simplification allows for the standard mathematical properties of the normal distribution to be applied in calculating the VaR estimate. This assumption contributes to the simplicity of this approach and should be used only for portfolios consisting of linear instruments (Dowd 1999). The major drawback of this method is the assumption of normality. Financial time series do not follow a normal distribution, and therefore, the assumption of normality can underestimate the potential risk especially when the tails of normal distribution are thinner than the return series distribution. This becomes a bigger issue when a portfolio includes non-linear instruments such as options. To overcome this limitation, a delta-normal approach is adopted, where the first order approximation of the returns series by liner approximation is used to calculate the VaR estimates. This improvement provided by the delta-normal approach comes at the expense of the simplicity of the variance-covariance approach.

8.3.3   Monte Carlo Simulation Method

The Monte Carlo simulation method is very similar to the historical simulation method. The main difference is that rather than using historical data for the simulation process, in the Monte Carlo simulation method, random numbers are generated from a statistical distribution that fits the return distribution. These random numbers are then used to calculate the hypothetical profit and loss of a portfolio. This can be used to determine the distribution of the portfolio profit and loss allowing for the calculation of the VaR measure.

The Monte Carlo simulation method comprises the following steps:

- Identify the market factors relevant to the portfolio in question.
- Determine a distribution that fits the returns series.
- Generate N random hypothetical values from the distribution.
- Subject the portfolio to the N random movements, by calculating the profit and loss of the portfolio using the changes in the market for each of the N random values.
- Order the portfolio profit and losses in descending order.
- Based on the desired confidence interval, select the appropriate value.

The Monte Carlo simulation is the most popular method for VaR calculation, but it is the most challenging to implement (Dowd 1999). The advantage of using this method is that it does not rely on the normality assumption of the return series and it can cater for non-linear instruments such as options. This method is also capable of generating the entire distribution of the underlying asset; therefore, it can be used to calculate losses in excess of VaR interval. The biggest challenge in implementing the Monte Carlo simulation method is the computational time required to generate reliable results. Implementing this method for portfolios with a large number of assets can become very expensive. Given the technological advances in computing processing power and the continuous drop computer hardware, this limitation will become less of an issue in the near future (Jorion 2001). A potential drawback of this method arises from *model risk* which is caused by making the wrong assumption about the pricing model and/or the underlying asset generating stochastic process. This wrong assumption can lead to undesirable outcomes, so the modeller requires subject matter expertise to derive the correct modelling technique. Also, explaining the results generated by this method to senior management would be very challenging.

8.4 Modelling Value-at-Risk with Neural Networks

VaR at risk models are used to obtain insight into a portfolio's risk exposure. Ascertaining this risk exposure requires a model that is capable of capturing key behaviours of all the instruments in the portfolio. Neural networks have been used with the aim of improving on the VaR models by utilising the neural networks' flexibility and modelling capabilities. Below is an overview of the research conducted in this domain.

(Locarek-Junge and Prinzler 1998) apply MDNs to approximate the distribution of the underlying returns series. The MDN's ability to approximate the underlying asset distribution can prove a powerful tool in estimating and predicting the VaR of a portfolio. The MDN was trained by minimising the negative log likelihood loss function using a back-propagation algorithm. The MDN is trained using 5-day and 30-day moving averages and moving standard deviations of the returns. The data series used for this study is the Deutsche Mark/US-Dollar exchange rates from September 1985 to August 1994. The first 2000 observations are categorised for the training set; the following 250 observations are used for testing. The number of hidden units and number of Gaussians were varied and the model with the best fit was selected. The VaR estimates were generated using Monte Carlo simulation with 10,000 simulation paths to approximate the distributions of the returns then calculating the 95 and 99% VaR estimates. The results show the MDN with one Gaussian has a similar performance to that of the Risk Metrics model, since both are autoregressive models and both assume normality in the underlying returns. However, once the number of Gaussians increases, a drop in the number of VaR exceedances is observed. The results show the MDN to have superior performance to that of the Risk Metric model based on the very basic testing criteria. Ideally, the MDN models would be back-tested using advanced methods. The other drawback of this study is that the design of the MDN is not consistent with the time series attributes, which potentially may lead the MDN to miss key behaviours such as the volatility feedback effect in the underlying data.

In (Miazhynskaia et al. 2003), linear and non-linear MDNs are applied to estimate and forecast the conditional distributions. These forecasts are then used to calculate the VaR measures for three different time series: Dow Jones Industrial average (DJIA), FTSE 100 traded on the London Stock Exchange and the Japan Index NIKKEI 225. The MDN is benchmarked against the GARCH model using Gaussian and Student-t distributions for each model. Each data series consisted of 13 years of observations from 1985 to 1997. The data series was divided into 500 days for training, followed by 100 days for validation and remaining 100 days for testing. All models were optimised by minimising the negative log-likelihood loss function. The MDN hidden units were set to three hidden units. The models were evaluated using five criteria which are: correctly estimating the VaR levels, Basle traffic light test, proportion failures test, conditional coverage test and lost interest yield. The results produced a different ranking for each model across different series. There was no significant advantage in using non-linear over linear models for the data series except for DJIA where linear models performed better than did the other models. The overall performance across all measures in the results support the non-Gaussian, which is understandable since the non-Gaussian models assign more mass in the tails therefore are able to cater for extreme movements in the underlying data series. The MDN performance varied across the time series which is a consequence of using one neural network structure to model different time series. Ideally, the MDN should be customised for each series by choosing the appropriate number of hidden units, number of Gaussians and the length of the training set.

In (Wu et al. 2005) quantile regression neural networks and support vector regression (SVR) are applied to VaR forecast. This technique aims at mapping the quantile to the return directly. Therefore, once the network is optimised, the return can be deduced from the neural network for any quantile. Both models were compared to the Monte Carlo Simulation VaR method. The composite index in the Shanghai Stock exchange data for the period December 1991 to February 2002 is used for optimisation of the models. The results demonstrate the superiority of the neural network and support vector machine compared to the Monte Carlo method; this can be explained by the normality assumption in the Monte Carlo simulation. The neural network is able to calculate the VaR with high confidence levels whereas the SVR is able to calculate VaR measures with higher accuracy; however, the SVR underestimates VaR with less probability. The drawback of this method is its ability to capture the volatility clustering in the underlying returns. Also, using extra key inputs such as volatility can contribute to the accuracy of the model.

(Dunis et al.) study Higher Order Neural Networks (HONNs) by benchmarking them against other models such as different neural network designs, Extreme Value Theory, ARMA-GARCH(1,1), Riskmetrics volatility models and hybrid neural networks. The models are evaluated based on the out-of-sample forecasting performance for computing the VaR measures. The data series used in this study is the Brent oil and Gold bullion for the period of April 2002 to March 2008. The data set was split into a training set of 1045 observation for training, 260 for validation and 261 for out-of-sample testing. The 10 past lag returns were chosen as inputs for the neural networks. For the MLP-Riskmetrics and HONN-Riskmetrics model, the historical volatilities of the past 10 lag returns were included as inputs. The results show the hybrid HONNs-Riskmetrics model has superior performance, with all neural network models demonstrating acceptable performance compared to the other statistical models. The drawback of this research is that the neural network design and training method was not explained: it was not clear if the neural network was trained to forecast returns or volatility. Also, the process of converting the neural networks forecasts to VaR measures was not covered.

In Chen et al. (2009), a back propagation neural network is used to predict the one-step-ahead conditional volatility. The conditional volatility is derived by estimating the conditional second moment and subtracting the squared neural network estimate of the expected return. Once the one-day-ahead conditional volatility is calculated, the VaR measure is then calculated by simulating the return distribution using Monte Carlo simulation. The data series used is 913 log returns of the Hang Seng Index from May 2003 to December 2006. The neural network inputs are chosen to be the last six lags which are based on autocorrelation analysis. The neural network performance is compared to the ARMA-GARCH model. The results show that the ARMA-GARCH model is accepted at the 97.5 and 95% confidence levels. However, the neural network is accepted on all levels. The limitation in this study is that the neural network training and model selection are not discussed. The network was trained on only 500 data points; this could cause issues for GARCH and neural network models.

8.5 Value-at-Risk (VaR)—Future Work

Many variations of the VaR models have been developed to increase the reliability and accuracy of the VaR measures. Neural networks have been proposed as an alternative to many econometric and statistical methods. Neural networks are a suitable choice for VaR models since they cater for nonlinear multivariate functions which are ideal for modelling VaR, since the portfolio can contain a wide range of assets that can have linear or non-linear payoffs. This is not a straightforward task, as neural networks have to be designed appropriately by selecting the right inputs, network architecture and training data. In Appendix B, a review of neural network application in VaR shows limited success in this domain. The weak performance of neural networks in modelling VaR can be attributed to a variety of factors. However, the key weakness is the ability of neural networks to forecast extreme values. This issue exists with statistical models which are compensated by using fat tail distributions. This is also dependent on the ability to capture the stylised facts of the underlying asset such as heteroscedasticity. This is a cumbersome task due to the lack of observation in the tails. Most models compensate for this by using fatter tail distribution rather than a normal distribution. This approach can have side effects as the model is forced to comply with the underlying asset, which can lead to over-estimation of VaR and therefore, cause more capital to be consumed for risk management purposes.

Neural networks can be utilised for modelling VaR as they have been successful in forecasting volatility and pricing options. Given the nature of this problem, we propose the MDN to address this issue rather than feedforward networks. The MDN will be able to capture the dynamics of the underlying asset while providing good approximation of the profit and loss distribution. Also, the MDN is able to cater for different stylised facts including heavy tails. Our proposed approach differs from other studies as our model captures the profit and loss distribution with the emphasis on learning the quantile value. This method will allow the MDN to infer VaR measures for any quantile value.

Input selection:

The inputs to the MDN are the quantile of the return, time index, volatility, and the target is the return value. To obtain the quantile we sort the returns from largest to smallest. Then we assign each return a number value between 0 and 1. So if the total number of returns is 100, we give the top return 1 and the next will be 1–0.001 etc.... Due to the absence of the return in the input, the time index is necessary to preserve the time dependency and sequencing in the underlying asset. The volatility parameter allows the MDN to feedback the volatility which allows it to model key stylised facts such as volatility clustering. The return value is used as the target since the MDN will require a mapping between the inputs to the return.

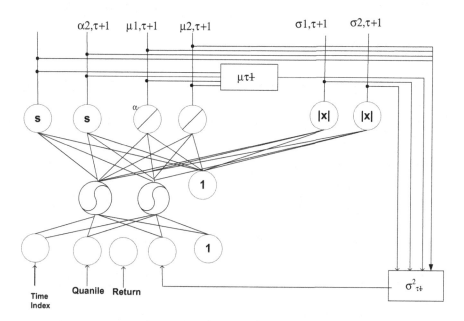

Fig. 8.2 MDN design for forecasting VaR

MDN Structure:

The number of output units and input units would depend on the time series and the training samples used. The MDN would need to be designed according to the guidelines highlighted in Chap. 4. A recurrent MDN structure is utilised to capture the dynamics of the time series as shown in Fig. 8.2.

If we define VaR to be as follows:

$$VaR^{\alpha}_{t+1} = \mu_{t+1}T - \phi^{-1}(\alpha)\sigma^2_{t+1}\sqrt{T} \tag{8.6}$$

Where μ_{t+1} is the expected return which can be calculated using ARIMA model, T is the time horizon measured in days, $\phi^{-1}(\alpha)$ is the α-quantile derived from the profit and loss distribution.

The MDN can be utilised to provide accurate estimation for $\phi^{-1}(\alpha)$, overcoming the assumption of normality. The advantages of this model over the models studied literature, is that it is directly modelling return quantile at the same time preserving the time dependency and dynamics of the underlying asset. One method will be considered is data boosting, which is achieved by generating more observation synthetically at the extreme values by preserving the statistical attributes of the data since we have the quantile as an input to the MDN. So by increasing the number of the observations at the tails, this can potentially increase the forecasting accuracy of the model.

Chapter 9
Conclusion and Discussion

Market risk is defined as the potential loss of an asset value due to movements in market factors. These markets factors could be anything that impact on the current or future value of an asset. The most important factor that has attracted significant attention in the literature is the asset volatility. Measuring the asset volatility is cumbersome since it is not directly measurable in the market, which then reduces it to a filtering and approximation task. To further complicate this problem, the volatility exhibits complex behaviours that must be understood and measured accordingly to ensure accurate forecasting and evaluation of the future asset value. Therefore, volatility modelling is at the heart of market risk. The success of the volatility models is very much reliant on their ability to capture the stylised facts of the underlying asset time series. In today's markets, many instruments are traded on the world market for many different reasons. These instruments can be used as investment or risk mitigation tools. However, if the instruments are not applied properly, substantial loss may be incurred. To avoid this potential risk, pricing models such as BSOPM have been developed to provide a fair value based on the current market conditions. In addition to evaluating the instrument, the option pricing models are also used for sensitivity analysis. For instance, ascertaining the sensitivity of the instrument to changes in the underlying stock value is useful in a delta hedging scenario. In this book, we study option pricing since options are the most common instruments traded on the world markets. The process of evaluating such instruments involves many variables such as interest rate, stock price, volatility etc. The pricing model needs to cater for the dynamic behaviour of the underlying asset and its evolution through time, in addition to the other factors that influence the option price. Therefore, the volatility plays a major role in the accuracy of the model to price the options. Some models use the historical volatility in the option formula, which affects the model's accuracy as the volatility is known not to be constant over time. This is one of the criticisms of the BSOPM which can be demonstrated by plotting the implied volatility against the strike price. For this reason, researchers had to formulate complex models such as the GOPM to address such issues. On a larger scale, it is cumbersome to analyse the potential market risk

© Springer International Publishing AG 2017

149

F. Mostafa et al., *Computational Intelligence Applications to Option Pricing, Volatility Forecasting and Value at Risk*, Studies in Computational Intelligence 697, DOI 10.1007/978-3-319-51668-4_9

of a portfolio which can contain many different instruments and assets. The overall profit and loss of the portfolio is a result of the interaction of all the portfolio components. To further complicate this analysis, the risk component in the portfolio can have linear and non-linear pay-offs, which means that the overall risk is not simply the sum of all risk components. The VaR model simplifies this issue by allowing the total risk value of the portfolio to be presented as a single number. This is calculated using the predictive distribution of future returns and taking a quantile value of the distribution. So, modelling the distribution of the portfolio returns is essential, although the distribution is generally assumed to be normal for simplicity.

Market risk modelling has become a hot topic in recent years due to the volatile market conditions. Investors are very much dependent on the risk models to understand and analyse their risk exposure. Market risk models generally work well in normal market conditions. However, they fail in extreme market conditions when they are most needed. This is a drawback of most statistical models where a prior assumption has to be made to allow for the derivation of the model. The model also contains a fixed number of variables which means it is limited accordingly. The modelling of financial time series would require a comprehensive model that is adaptable and flexible to avoid the strict assumption about the underlying data. Neural networks are well known for their flexibility and approximation capabilities. Also, neural networks are data-driven, so no prior assumption about the data is needed. Specifically for market risk modelling, neural networks are most suitable as they can capture complex nonlinearities and subtle relationships in the data. For this reason, neural networks have been applied to many market risk problems with the aim of overcoming the limitations of the statistical models.

In Chaps. 2 and 3, we provided a detailed review of the models applied in market risk mainly in volatility forecasting, option pricing and VaR modelling. We reviewed the neural network literature in this domain in Chap. 4 and discussed the limitations of each approach. This was then used as the foundation for the rest of the book. In Chap. 5, we provide an overview of the solutions for the problems addressed in this book. We discuss below the findings of Chaps. 5 and 6 and a recommendation for VaR modelling is presented based on the outcomes and findings of this book.

9.1   Volatility Forecasting

Financial time series are studied in terms of asset returns rather than asset prices as the return series can explain the investor's position, and can also identify potential risks or investment opportunities. To capture this vital information, statistical models are used to evaluate and forecast the asset price using the historical return series. Traditionally, return series were modelled using ARIMA models, which can explain some of the behaviours in the return series. The main drawback of such models is the assumption of a constant variance; hence, the models are not able to cater for key stylised facts which include heteroscedasticity, volatility clustering and persistence of the volatility episodes. If we adopted Wold's decomposition theorem, the return series can be split into deterministic and non-deterministic components. The deterministic component is the conditional mean (expectation) which can be captured in the ARIMA model. The non-deterministic component is the residual value that is considered to be a random variable that contains vital information such as the volatility of the return series.

The volatility is not directly observable, so this transforms the problem to a filtering and approximation problem. There have been many methods proposed for calculating the underlying asset volatility which range from calculating the variance using historical data to more complicated methods such as exponential weighting moving average. These approaches have been adopted for their simplicity. However, they assume the variance is homoscedastic which is contradictory to the stylised facts observed in the majority of return series. The volatility of the underlying asset series exhibits complex behaviours such as volatility clustering and heteroscedasticity. This issue was addressed by the ARCH model, where the initial formulation catered for some key volatility attributes such as volatility clustering and persistence. Many versions of the ARCH model have emerged to address its limitations. For instance, the GARCH model reduces the number of lagged variables required in the variance equation. The EGARCH model caters for the leverage effects in the returns series. To date, no single model caters for every aspect of the underlying asset series. For this reason, we see many variations of the ARCH model that are designed to address different aspects of the underlying return series.

Neural networks have been applied in the forecasting domain in the hope of overcoming the limitations of the statistical models. The neural networks are non-linear models that are capable of approximating any continuous function with a certain degree of accuracy (Hornik et al. 1989; Hornik 1993). These generalisation capabilities, with the flexibility of the neural network model, position it as the ideal forecasting tool. In the literature, the majority of the volatility forecasting papers use a hybrid approach where the neural network is combined with the statistical models leveraging of each other to provide accurate forecasts. The reasoning behind this approach is that the standard feedforward neural network is limited in modelling the conditional volatility. That is, the loss function used in the optimisation process such as the sum of error square (SSE) has a constant variance with respect to the input variables. This means the standard feedforward neural network is not capable of modelling heteroscedasticity in the data which will impact on the accuracy of volatility forecasting. This limitation was overcome by MDN where the variance is a function of the input variables; therefore, the variance can change over the training population preserving the heteroscedasticity in the data. The flexibility of the MDN allows it to be designed to mimic the GARCH model. This is achieved by allowing the feeding back of the variance as an input. In analysing the literature, we find discrepancies in the results reported for volatility models and neural networks alike. These discrepancies are discussed earlier and further analysed in this chapter.

In our analysis of the volatility forecasting literature, we observed common oversights and practices that are widely adopted for time series and neural network models alike. One such oversight is that the number of observations in the training set is often chosen in an ad hoc manner. This practice can cause misleading results as over-fitting and under-fitting of the model parameters are directly related to the number of observations used in the training set. Using too few observations in the training data set can lead to under-fitting which can cause instability of the model and poor forecasting performance. Also, using too many observations can cause over-fitting and lead to memorisation or loss of generalisation. This can be seen in the results shown in Sect. 6.7, where the MDN forecasts of the NAB and QAN provide much better forecasts using 750 observations; whereas, ANZ and BHP provide better forecasts with 1000 observation. It is interesting to see that the GARCH models also displayed similar behaviour where the GARCH model displayed superior performance for the QAN daily forecast using 1000 returns, whereas the EGARCH model was superior to the GARCH using the 750 returns. A training set that spans a long period of time can potentially contain past behaviours that are no longer valid in today's market. Also, a small training set would not contain enough information for the model to make an accurate prediction of the future.

Over-fitting and under-fitting are also made possible by the number of parameters in the model. Neural networks are more prone to this issue due to the high number of weights connections. In the literature, the number of hidden units is generally chosen based on the individual past experience. In Sect. 6.7, the impact of using the wrong number of hidden units is highlighted in the results. It is interesting to see the performance of the MDN varying substantially for the same dataset as the number of hidden units and Gaussians vary. The optimal number of hidden units and Gaussians differ across different time series according to the number of observations in the training dataset. Therefore, mis-specifying the number of hidden units and Gaussians in the MDN has a direct impact on the forecasting performance of the MDN in the short and long term.

Several alternative models are often proposed to explain the same data, where objective criteria are needed to choose the most suitable model. While adding extra parameters to a model is often desirable, the increased complexity comes at a cost. The criteria for selecting the most suitable model must weigh the trade-off between increased information captured in the model by increased parameters and decreased reliability caused by over-fitting. This has been formulated in the famous Akaike Information Criterion (AIC). The decrease in the likelihood function should be sufficient to offset the use of the extra parameters in the model. There have been many theories developed to determine the optimal neural network size, such as the NIC (network information criteria) (Amirikian and Nishimura 1994) which is a generalisation of the AIC. In this book, the optimal model was selected based on the error measures explained in Sect. 6.4. It was interesting to see the optimal model differs across different time series and differed for different forecast horizons. This suggests that the MDN is still not completely optimised. The MDN could be finetuned by using pruning algorithms such as those in Reed (1993). Pruning a neural network reduces the effect of spurious data, which will improve the accuracy of the forecast. The generalisation power of the MDN and its ability to capture the underlying asset dynamics of the returns series such as high volatility persistence makes it a better time series model than the traditional GARCH models. As demonstrated by Diebold and Pauly (1987) and Lamoureux and Lastrapes (1990), high volatility persistence in the GARCH model could be due to structural changes in the variance process. The MDN has the capability of capturing such dynamic behaviour, which is evident in its superior performance compared to that of the GARCH models.

Computational issues were observed in the initial stages of this research. When the parameters for the GARCH model were estimated using Matlab and Eviews, the results were not consistent and in some cases they were strikingly different. This behaviour could be explained by the difference in the way the variance is calculated (Bruno and De Bonis 2000). Also, when optimising a model using different algorithms within the same package, the estimated parameter values were completely different. This can be expected due to different initial conditions terminating criteria used by the algorithms. In some cases, when using the BHHH algorithm (especially when using 750 returns), the training process was terminated prematurely. A different algorithm had to be used, such as the Marquardt algorithm to overcome this convergence issue. These computational issues are also present when training neural networks. Lawrence and Giles (2000) showed that using the conjugate gradient algorithm to train an MLP can lead to worse generalisation because they can be more prone to creating varying degrees of over-fitting in different regions of the input space.

In the light of the arguments presented above, all models could be further finetuned. Also, the results explain the mixed results seen in the literature. The purpose of this study is to evaluate different volatility model performances with respect to different time series. Based on the statistical criteria used, it can be seen that MDN is superior to the traditional GARCH models. The statistical evaluation criteria used have no economic meaning. Also, they do not incorporate uncertainty, due to parameter estimations. It is recommended that the model forecast should be evaluated according to economic loss rather than statistical loss as suggested in Lopez (2001).

9.2 Option Pricing and Hedging

Options are the most common instruments traded on the world markets. What makes options so popular amongst investors is that they can be used as an investment or a risk management tool. However, they can be a source of risk if not applied appropriately as the investor can be exposed to extreme losses. Investors are heavily dependent on option pricing models to provide a fair evaluation of the option. The wide success of this model was due to its ability to produce an option price instantaneously. The formulation of BSOPM was based on major assumptions which allowed for the derivation of this model (see Sect. 7.1.2). These assumptions significantly impacted the model pricing capabilities which are visible in the models pricing biases. The constant volatility assumption has captured the attention of researchers, as it contradicts the volatility behaviour observed in the underlying asset. Also, it was realised when the volatility was backed out from the BSOPM for different strike prices, the implied volatility displays a U-shape pattern (volatility smile) when plotted against moneyness of the option. Nevertheless, implied volatility is derived directly from the options data, which contains vital information as it provides a foreword looking view on how the market is anticipating movements in the underlying asset. This is a different view to the conditional volatility calculated from the stock returns which is a backward view. The other interesting feature of the implied volatility is the deformation of the volatility surface through time caused by the changes in the implied volatility levels. Many models have emerged to overcome the limitations of the BSOPM. In this research, we select the GOPM as it allows for the volatility to follow a GARCH specification. This specification overcomes the constant volatility assumptions and at the same time caters for key stylised facts in the underlying returns series. This model has displayed superior improvements over other pricing models. The major shortcoming for this model is the absence of a closed form solution. This means the model parameters have to be optimised using the Monte Carlo simulation, which is a computationally intensive process that limits its alibility to be applied to real-life problems.

Neural networks have been applied heavily to option pricing problems. The majority of research papers rely on mapping inputs similar to the BSOPM to the option price. This approach displays some improvements in pricing accuracy over the BSOPM. This is not surprising since neural networks are capable of learning the relationships between the input variables. Yet there has been no attempt to evaluate the neural networks' performance against more advanced models. The reason for this is that the pricing capabilities of the neural networks are inferior to those of more advanced pricing model such as the GOPM. In Sect. 4.2, we explain that this behaviour is caused by the approach adopted in training the neural networks. Given the neural networks' approximation capabilities, they should be able to learn key behaviours in the option data to provide accurate option pricing that is comparable to any statistical model such as the GOPM. To overcome this limitation, we propose a new method that utilises the implied volatility rather than the option price. This will allow the neural network to learn the implied volatility surface across all maturities and strike prices. Once the neural network is optimised, it will be able to produce an implied volatility for any maturity and strike. The implied volatility value can then be converted to an option price by using the BSOPM. To compare the performance of our proposed method, we benchmark it against another neural network model that is trained on the option price. We also do the same analysis for the BSOPM and the GOPM. This will allow us to demonstrate the effectiveness of our new method with respect to the other models. This is the first study to analyse the option pricing performance of a neural network in comparison to the GOPM. In Chap. 6, we conduct the experiments using FTSE 100 call options. Our new method

(NNiv) on average displays superior pricing compared to that of all other models. The exception is for the DITM and DOTM options. This is due to the number of options that fall into this category which could be caused by the data cleansing exercise. These are very impressive results since, for the first time, the neural networks are capable of pricing options with higher accuracy than do advanced stochastic option pricing models. The results in Chap. 6 are very encouraging, and to further analyse the capabilities of our new method, we conduct a delta hedging exercise. The aim of the delta hedging exercise is to demonstrate the completeness of the model. That is, a correctly specified option pricing model should be able to predict future movement in the underlying asset. The hedging results are somewhat surprising as the GOPM performed the worst and the NNiv was on par with BSOPM. The NNiv method performed very well on the short maturities and in-the-money options. In terms of the GOPM hedging performance, the model parameters were estimated only on a single day; therefore, the model's parameters would not contain the required information to make a prediction on the underlying asset movements. This also is evident in Fig. 6.2 and Fig. 6.3, where the implied volatility produced by the GOPM deforms with time. It would be interesting to see if this still holds when the GOPM is optimised on a larger data set. The neural network model was trained on a larger data set and by having a higher number of variables (i.e. weights), the neural network would be able to capture more of the underlying asset dynamics. In Fig. 6.4 and Fig. 6.5, the implied volatility does not display any deformation through time which shows the NNiv model is able to predict the implied volatility well into the future. The delta hedging exercise is not practical, since the option moneyness with change as the underlying asset price changes. So the investor would not necessarily hold on to the option until expiry.

9.3 Recapitulation

In this book, we investigate the application of neural networks to ascertain market risk. This area of research has seen increased interest in recent years due to the increased volatilities and uncertainty in the global financial markets. Investment firms are continuously seeking new models which will enable them to understand and control their risk exposure. In this research, we address the poor results produced by neural networks as reported in the literature and provide explanations for these results. We make a direct contribution to the field of artificial intelligence by providing explanations for the variation of the results shown in literature for different artificial intelligence models. Also, we demonstrate the capabilities of such techniques in overcoming computational and assumptions issues inherited from statistical models. Specifically, we demonstrate the importance of neural networks techniques in addressing the limitation of financial risk models.

Firstly, we provided a detailed literature review in Chaps. 2 and 3. The review consists of t financial modelling techniques where financial time series modelling, option pricing, hedging and VaR models are reviewed. In each of these areas, we elaborate on the strengths and weaknesses of the models under study relative to the reported results of the relevant study. The second main area covered in this chapter is neural networks. We reviewed the neural network models that have been adopted for financial modelling in Chap. 4, which are mainly back-propagating MLP, MDN and RBF networks. We then present a comprehensive review of neural network research conducted in conditional volatility forecasting, option pricing, hedging and VaR. In this review, we highlight the strengths and weaknesses by analysing the methods applied and results for each study. This review is an accomplishment as it can serve as a main reference for future research. Moreover, this detailed analysis allowed us to formulate the objectives of this book by highlighting the inconsistencies in the literature which have been the main motivation for this book.

In Chap. 5, a clear and precise problem definition is formulated. This is achieved by defining the terms and concepts to be used throughout the book. This includes the problem definition and the choice of methodology to be followed throughout this book. In Chap. 6, we provide a solution overview for each of the main problems that are addressed in this research which include volatility forecasting, option pricing and hedging. The conditional volatility modelling solution addresses the modelling issues for statistical and neural networks models which are typically overlooked, and the impact that this oversight can potentially have on the stability and forecasting performance of the models. To further elaborate on this, we propose testing the models against different time series to evaluate the behaviour of the models. For option pricing, we note that neural networks results are very poor in this domain. The potential cause of this weak pricing performance is explained and an alternative method for pricing option is suggested. To evaluate the effectiveness of this new method, it is proposed that it be evaluated against the common approach used in neural network literature and the GOPM. This will be the first study of its kind to compare option pricing of neural networks with advanced stochastic pricing models. To further analyse this new method, we adopt delta hedging scenario to evaluate the competence of our proposed model with that of the competing models.

In Chap. 6, we also described extensive experiments on volatility modelling conducted. The results obtained from these experiment highlights and explain the results reported in the literature. Most significantly, the results demonstrate the importance of selecting the correct model structure for time series understudy, and the impact of common oversights such as the number of observations in the training set length which are often neglected and can impact on the performance of the time series and neural network models alike. The results also highlight the flexibility of the neural networks and ability to capture complex behaviour in the underlying data. The neural networks on overage outperformed the GARCH and EGARCH models subject the correct model specification. The results in this chapter can be used for future research as a guide for designing a neural network that meet the objective target.

In Chap. 7, the option pricing experiment was conducted. Our proposed model deviates from the approach adopted in the literature, since our model is trained on the implied volatility rather than option price. The results obtained from experiments show the superior option pricing performance of our model relative to other neural network models as well as the BSOPM and GOPM. These results are very impressive as this is the first time that neural networks have been able to outperform advanced option pricing models such as GOPM. To further verify and test our model, we performed a delta hedging exercise. The hedging results were somewhat surprising, as our proposed model was on par with the BSOPM, and GOPM had the worst performance. The results presented in this chapter demonstrate the effectiveness and power of neural networks to learn and capture complex patterns in the data. In Chap. 7, the results for the experiment are further explained and the contributions of each chapter highlighted. In Chap. 8 we examined the Value at Risk problem.

Neural network are known to be universal approximators that can approximate any continuous function with certain accuracy. This theory seems to fail when applied to financial problems such as volatility forecasting and option pricing. In this book, we show that the lack of success of neural networks in modelling market risk aspects is due to the neural network design and training. We have demonstrated in this book the superiority of neural network over statistical models as long as the neural network model is designed and trained appropriately. The results obtained in this book serve to motivate the further pursuit of neural networks as a risk management tool that can address the many limitations of the current statistical models.

9.4 Contributions of This Research

In this book we make the following contributions:

- This book directly contributes to the field of artificial intelligence by providing explanations for the variation of the results shown in literature for different artificial intelligence financial time series forecasting and option pricing models. We use life financial data to demonstrate the capabilities of such techniques in overcoming computational and assumptions issues inherited from statistical models.

- Specifically, we demonstrate the importance of neural networks techniques in addressing the limitation of financial risk models. We contribute to the field of neural networks not only by demonstrating the superiority of the neural networks over the state of the art statistical models, but we also emphasis the selection and the design of the right model for the task. This can be seen with the mixture density networks volatility forecasting accuracy varying dramatically with different model configurations. This is also apparent when using the backpropagating neural networks to price options using the option price and implied volatility as the target output.

- In this book we demonstrate the superiority of the mixture density networks in forecasting volatility over GARCH and EGARCH models. We demonstrate the impact of using the correct model specification and the length of the training data set on the performance of the mixture density networks and time series models alike. The results obtained here can be utilised as a guide in designing neural networks forecasting models to overcome the common assumptions adopted in volatility forecasting literature.

- To our knowledge (at the time of writing this book), this is the first in-depth study to evaluate the option pricing and hedging capabilities of neural networks against the GARCH option pricing model. The neural network model proposed in this book was able to outperform the Black-Scholes and GARCH option pricing mode. We provide explanations to the downfalls of the option pricing models used in literature and propose a new method which allows the neural network to learn the implied volatility surface rather than the option price. This is a significant contribution as for the first time demonstrate the capabilities of the neural networks to outperform advanced option pricing models.

- The results obtained in this research contribute to the market risk literature by demonstrating the superiority of the neural networks over state of the art forecasting and option pricing models. The techniques and models used in this research can therefore be applied directly to real life trading and can also be extended to include other risk models such as Value at Risk models.

Appendix A
Detailed Results

The results below have been summarised and presented in Chaps. 6 and 7. The full set of results is displayed below. The numbers shown in bold in the tables below indicate the best performing model.

A.1 One Day Forecast Results

See Tables A.1, A.2, A.3 and A.4.

Table A.1 1-day forecast for 1000 returns

	BHP		Qan		NAB		ANZ	
	NMAE	HR	NMAE	HR	NMAE	HR	NMAE	HR
GARCH	0.694	74	**0.914**	68	0.790	70	**0.796**	**73**
EGARCH	**0.667**	74	0.968	65	**0.745**	74	0.796	70
LF EGARCH	**2.190**		2.466		**2.754**		**2.545**	
LF GARCH	2.014		**2.379**		2.788		2.578	

Table A.2 1-day forecast for 750 returns

	BHP		Qan		NAB		ANZ	
	NMAE	HR	NMAE	HR	NMAE	HR	NMAE	HR
GARCH	0.893	63	0.746	72	**0.678**	**74**	0.872	68
EGARCH	**0.847**	65	**0.737**	72	0.705	74	**0.842**	69
LF EGARCH	**3.007**		**2.524**		2.480		**2.911**	
LF GARCH	3.002		2.520		**2.666**		2.909	

© Springer International Publishing AG 2017
F. Mostafa et al., *Computational Intelligence Applications to Option Pricing,
Volatility Forecasting and Value at Risk*, Studies in Computational
Intelligence 697, DOI 10.1007/978-3-319-51668-4

Table A.3 MDN 1-day-ahead forecast for 1000 returns

	BHP			ANZ			NAB			QAN		
	NMAE	HR	LLF	NMAE	HR	LLF	NMAE	HR	LLF	NMAE	HR	LLF
2G2H1k	0.685	72	1.874	0.752	75	1.263	0.740	74	1.291	1.007	65	1.216
2G3H1k	0.688	72	1.777	0.753	75	1.263	0.740	74	1.291	1.007	64	1.229
2G4H1k	0.685	72	1.926	0.754	75	1.265	0.739	75	1.240	0.995	64	1.279
2G5H1k	0.686	72	1.868	0.741	76	1.161	0.740	75	1.294	1.007	64	1.238
2G8H1k	0.689	72	1.652	0.753	75	1.263	0.740	74	1.291	1.008	63	1.242
3G2H1k	*0.662*	*74*	*2.261*	0.762	75	1.264	*0.720*	*75*	*1.340*	0.992	63	1.237
3G3H1k	0.679	72	2.049	0.761	75	1.264	0.719	75	1.330	0.995	64	1.189
3G4H1k	0.684	72	1.926	*0.753*	*75*	*1.282*	0.722	74	1.298	*0.991*	*64*	*1.342*
3G5H1k	0.687	74	2.187	0.761	75	1.264	0.720	75	1.340	0.991	64	1.250
3G8H1k	0.683	72	1.924	0.761	75	1.263	0.720	75	1.330	0.990	64	1.212

Table A.4 MDN 1-day forecast for 750 returns

	BHP			ANZ			NAB			QAN		
	NMAE	HR	LLF	NMAE	HR	LLF	NMAE	HR	LLF	NMAE	HR	LLF
2G2H750	0.741	73	3.880	0.854	65	1.346	0.720	72	1.291	0.723	76	1.216
2G3H750	0.753	68	3.907	0.905	65	1.297	0.722	73	1.291	0.726	76	1.229
2G4H750	0.936	63	3.060	0.918	65	1.298	0.722	73	1.291	0.722	76	1.221
2G5H750	0.735	71	3.884	0.854	65	1.347	0.722	74	1.294	0.732	76	1.238
2G8H750	0.745	72	3.886	0.904	65	1.302	0.722	73	1.291	0.733	76	1.242
3G2H750	0.749	69	3.896	0.856	65	1.367	*0.722*	*73*	*1.340*	0.730	76	1.237
3G3H750	0.738	73	3.878	0.858	65	1.368	0.725	72	1.330	0.710	76	1.189
3G4H750	1.088	60	3.009	0.857	65	1.368	0.723	73	1.338	0.739	76	1.250
3G5H750	*0.742*	*72*	*3.906*	*0.857*	*65*	*1.369*	0.723	73	1.340	*0.738*	*76*	*1.250*
3G1H750	0.727	70	3.846	0.857	65	1.370	0.725	72	1.330	0.722	76	1.212

A.2   10 Day Forecast Results

See Tables A.5 and A.6.

Table A.5 10-day forecast for 1000 returns

	BHP			ANZ			NAB			QAN		
	NMAE	HR	LLF	NMAE	HR	LLF	NMAE	HR	LLF	NMAE	HR	LLF
2G2H1k	*0.919*	*0.70*	*1.297*	0.584	0.80	2.384	0.569	0.80	2.024	0.748	0.80	3.007
2G3H1k	0.978	0.60	0.371	0.600	0.80	2.381	0.569	0.80	2.021	0.732	0.80	3.009
2G4H1k	0.971	0.60	0.593	0.537	0.90	2.028	0.569	0.80	2.026	0.746	0.80	3.008
2G5H1k	0.959	0.60	0.789	0.569	0.80	2.396	0.568	0.80	2.005	0.751	0.80	3.008
2G8H1k	0.963	0.60	0.747	*0.535*	*0.90*	*2.095*	0.566	0.80	1.986	0.765	0.80	3.006
3G2H1k	0.970	0.60	0.577	0.538	0.90	1.968	*0.571*	*0.80*	*2.051*	0.740	0.80	3.008
3G3H1k	0.974	0.60	0.513	0.636	0.80	2.353	0.569	0.80	2.032	*0.681*	*0.80*	*3.018*
3G4H1k	0.961	0.60	0.752	0.597	0.80	2.374	0.571	0.80	2.047	0.796	0.80	3.002
3G5H1k	0.950	0.60	0.979	0.557	0.80	2.391	0.566	0.80	1.987	0.749	0.80	3.008
3G8H1k	0.973	0.60	0.509	0.608	0.80	2.362	0.565	0.80	1.968	0.745	0.80	3.008

Table A.6 10-day forecast for 750 returns

	BHP			ANZ			NAB			QAN		
	NMAE	HR	LLF	NMAE	HR	LLF	NMAE	HR	LLF	NMAE	HR	LLF
2G2H750	0.756	0.60	2.495	0.853	0.60	2.998	0.790	0.80	3.224	0.601	0.80	2.844
2G3H750	0.757	0.60	3.571	0.828	0.60	3.346	0.776	0.80	3.631	0.597	0.80	3.553
2G4H750	0.752	0.60	3.541	0.839	0.60	3.337	0.794	0.80	3.617	0.602	0.80	3.532
2G5H750	**0.719**	**0.60**	**3.341**	0.821	0.60	3.351	0.797	0.80	3.615	**0.585**	**0.80**	**3.605**
2G8H750	0.741	0.50	3.485	0.831	0.60	3.344	**0.774**	**0.80**	**3.633**	0.600	0.80	3.539
3G2H750	0.746	0.60	3.509	0.834	0.60	3.342	0.783	0.80	3.626	0.599	0.80	3.548
3G3H750	0.749	0.60	3.527	**0.797**	**0.60**	**3.371**	0.792	0.80	3.618	0.602	0.80	3.531
3G4H750	0.732	0.50	3.440	0.822	0.60	3.350	0.795	0.80	3.616	0.602	0.80	3.535
3G5H750	0.757	0.60	3.572	0.804	0.60	3.365	0.785	0.80	3.624	0.603	0.80	3.529
3G1H750	0.758	0.60	3.575	0.812	0.60	3.358	0.786	0.80	3.623	0.603	0.80	3.527

A.3 30 Day Forecast Results

See Tables A.7 and A.8.

Table A.7 30-day forecast for 1000 returns

	BHP			ANZ			NAB			QAN		
	NMAE	HR	LLF	NMAE	HR	LLF	NMAE	HR	LLF	NMAE	HR	LLF
2G2H1k	*0.735*	*0.83*	*1.366*	0.945	0.70	2.432	0.675	0.83	2.365	0.924	0.63	2.839
2G3H1k	0.764	0.77	0.488	0.980	0.67	2.424	0.675	0.83	2.364	0.904	0.63	2.843
2G4H1k	0.760	0.77	0.658	*0.750*	*0.80*	*2.260*	0.675	0.83	2.367	0.921	0.63	2.840
2G5H1k	0.754	0.77	0.875	0.913	0.70	2.442	0.674	0.83	2.354	0.928	0.63	2.839
2G8H1k	0.756	0.77	0.822	0.750	0.80	2.289	0.673	0.83	2.340	0.944	0.63	2.836
3G2H1k	0.760	0.77	0.675	0.750	0.80	2.235	*0.676*	*0.83*	*2.384*	0.914	0.63	2.841
3G3H1k	0.762	0.77	0.599	1.065	0.67	2.398	0.675	0.83	2.371	*0.845*	*0.60*	*2.855*
3G4H1k	0.755	0.77	0.842	0.975	0.67	2.423	0.676	0.83	2.381	0.981	0.63	2.828
3G5H1k	0.750	0.77	1.018	0.887	0.70	2.445	0.673	0.83	2.341	0.926	0.63	2.839
3G1H1k	0.762	0.77	0.604	1.001	0.67	2.414	0.673	0.83	2.327	0.920	0.63	2.840

Table A. 8 30-day forecast for 750 returns

	BHP			ANZ			NAB			QAN		
	NMAE	HR	LLF	NMAE	HR	LLF	NMAE	HR	LLF	NMAE	HR	LLF
2G2H750	0.741	0.67	2.776	0.766	0.63	2.569	0.705	0.77	2.282	*0.597*	*0.80*	*2.365*
2G3H750	0.741	0.67	3.571	0.760	0.63	3.345	0.704	0.77	3.630	0.595	0.80	3.229
2G4H750	0.747	0.67	3.542	0.762	0.63	3.337	0.705	0.77	3.617	0.597	0.80	3.211
2G5H750	0.814	0.63	3.341	0.758	0.63	3.350	0.705	0.77	3.615	0.590	0.80	3.274
2G8H750	0.761	0.63	3.485	0.760	0.63	3.344	*0.704*	*0.77*	*3.633*	0.596	0.80	3.216
3G2H750	0.755	0.67	3.509	0.761	0.63	3.342	0.704	0.77	3.626	0.595	0.80	3.225
3G3H750	0.750	0.67	3.527	*0.752*	*0.63*	*3.369*	0.705	0.77	3.618	0.597	0.80	3.210
3G4H750	0.773	0.63	3.440	0.758	0.63	3.349	0.705	0.77	3.616	0.597	0.80	3.213
3G5H750	0.740	0.67	3.572	0.754	0.63	3.363	0.704	0.77	3.624	0.597	0.80	3.209
3G1H750	*0.740*	*0.67*	*3.575*	0.756	0.63	3.357	0.704	0.77	3.623	0.598	0.77	3.206

References

Aboura, S. 2005. GARCH option pricing under skew. *Journal: The ICFAI Journal of Applied Economics* 4 (6): 78–86.

Akgiray, V. 1989. Conditional heteroscedasticity in time series of stock returns: Evidence and forecasts. *The Journal of Business* 62 (1): 55–80.

Amilon, H. 2003. A neural network versus Black-Scholes: A comparison of pricing and hedging performances. *Journal of Forecasting* 22 (4): 317–335.

Amin, K.I., and V.K. Ng. 1993. Option valuation with systematic stochastic volatility. *Journal of Finance* 48 (3): 881–910.

Amirikian, B., and H. Nishimura. 1994. What size network is good for generalization of a specific task of interest? *Neural Networks* 7 (2): 321–329.

Ammann, M., and C. Reich. 2001. VaR for nonlinear financial instruments—Linear approximation or full Monte Carlo? *Financial Markets and Portfolio Management* 15 (3): 363–378.

Andersen, T.G., and T. Bollerslev. 1997. Intraday periodicity and volatility persistence in financial markets. *Journal of Empirical Finance* 4 (2–3): 115–158.

Andersen, T.G., and T. Bollerslev. 1998. Answering the skeptics: Yes, standard volatility models do provide accurate forecasts. *International Economic Review* 39 (4): 885–905.

Andersen, T.G., T. Bollerslev, et al. 2001. The distribution of realized stock return volatility. *Journal of Financial Economics* 61 (1): 43–76.

Asay, M., and C. Edelsburg. 1986. Can a dynamic strategy replicate the returns of an option? *Journal of Futures Markets* 6 (1): 63–70.

Baillie, R., and R.P. DeGennaro. 1990. Stock returns and volatility. *Journal of Financial & Quantitative Analysis* 25 (2).

Baillie, R.T., and T. Bollerslev. 1989. Common stochastic trends in a system of exchange rates. *Journal of Finance* 44 (1): 167–181.

Banerjee, A., J.J. Dolado, et al. 1986. Exploring equilibrium relationships in econometrics through static models: Some monte carlo evidence. *Oxford Bulletin of Economics and Statistics* 48 (3): 253–277.

Baum, E.B., and D. Haussler. 1989. What size net gives valid generalization? *Neural Computation* 1 (1): 151–160.

Bennell, J., and C. Sutcliffe. 2004. Black-Scholes versus artificial neural networks in pricing FTSE 100 options. *Intelligent Systems in Accounting, Finance & Management* 12 (4): 243–260.

Bera, A.K., and M.L. Higgins. 1993. ARCH models: Properties, estimation and testing. *Journal of Economic Surveys* 7 (4): 305–366.

Billingsley, P. 2008. *Probability and measure.* Wiley India Pvt. Ltd.

Bishop, C.M. 1996. *Mixture density networks.* Neural Computing Research Croup Report NCRG/4288. United Kingdom: Aston University.

Black, F. 1975. Fact and fantasy in the use of options. *Financial Analysts Journal* 31 (4): 36–72.

Black, F. 1976. The pricing of commodity contracts. *Journal of Financial Economics* 3 (1–2): 167–179.

© Springer International Publishing AG 2017

163

F. Mostafa et al., *Computational Intelligence Applications to Option Pricing, Volatility Forecasting and Value at Risk*, Studies in Computational Intelligence 697, DOI 10.1007/978-3-319-51668-4

Black, F., and M. Scholes. 1972. The valuation of option contracts and a test of market efficiency. *Journal of Finance* 27 (2): 399–417.

Black, F., and M. Scholes. 1973. The pricing of options and corporate liabilities. *The Journal of Political Economy* 81 (3): 637–654.

Blair, B.J., S.H. Poon, et al. 2001. Forecasting S&P 100 volatility: The incremental information content of implied volatilities and high-frequency index returns. *Journal of Econometrics* 105 (1): 5–26.

Bollerslev, T. 1986. Generalized autoregressive conditional heteroskedasticity. *Journal of econometrics* 31 (3): 307–327.

Bollerslev, T. 1987. A conditionally heteroskedastic time series model for speculative prices and rates of return. *The Review of Economics and Statistics* 542–547.

Bollerslev, T., and R.F. Engle. 1993. Common persistence in conditional variances. *Econometrica: Journal of the Econometric Society* 61 (1): 167–186.

Bollerslev, T.I.M. 1995. *Generalized autoregressive conditional heteroskedasticity.* ARCH: Selected Readings.

Boyle, P.P., and D. Emanuel. 1980. Discretely adjusted option hedges. *Journal of Financial Economics* 8 (3): 259–282.

Boyle, P.P., and T. Vorst. 1992. Option replication in discrete time with transaction costs. *Journal of Finance* 271–293.

Brennan, M.J. 1979. The pricing of contingent claims in discrete time models. *Journal of Finance* 53–68.

Bruno, G., and R. De Bonis. 2000. A comparative study of alternative econometric packages: An application to Italian deposit interest rates.

Burstein, F., and S. Gregor. 1999. The systems development or engineering approach to research in information systems: An action research perspective.

Campbell, J.Y., and L. Hentschel. 1992. No news is good news: An asymmetric model of changing volatility in stock returns. *Journal of Financial Economics* 31 (3): 281–318.

Canina, L., and S. Figlewski. 1993. The informational content of implied volatility. *Review of Financial Studies* 6 (3): 659.

Cao, C.Q., and R.S. Tsay. 1992. Nonlinear time series analysis of stock volatilities. *Journal of Applied Econometrics* 7 (S1): S165–S185.

Chakraborty, K., K. Mehrotra, et al. 1992. Forecasting the behavior of multivariate time series using neural networks. *Neural Networks* 5 (6): 961–970.

Chan, W.S., and H. Tong. 1986. On tests for non-linearity in time series analysis. *Journal of Forecasting* 5 (4): 217–228.

Chauvin, Y., and D.E. Rumelhart. 1995. *Backpropagation: Theory, architectures, and applications.* Lawrence Erlbaum.

Chen, X., K.K. Lai, et al. 2009. *A statistical neural network approach for value-at-risk analysis.* IEEE.

Chen, Y.T., and C.M. Kuan. 2002. Time irreversibility and EGARCH effects in US stock index returns. *Journal of Applied Econometrics* 17 (5): 565–578.

Christensen, B.J., and N.R. Prabhala. 1998. The relation between implied and realized volatility. *Journal of Financial Economics* 50 (2): 125–150.

Christie, A.A. 1982. The stochastic behavior of common stock variances: Value, leverage and interest rate effects. *Journal of Financial Economics* 10 (4): 407–432.

Christoffersen, P., and K. Jacobs. 2004. Which GARCH model for option valuation? *Management Science* 50 (9): 1204–1221.

Coleman, T.F., Y. Kim, et al. 2001. Dynamic hedging with a deterministic local volatility function model. *Journal of Risk* 4 (1): 63–89.

Cont, R., and J. da Fonseca. 2001. Deformation of implied volatility surfaces: An empirical analysis. *Empirical approaches to financial fluctuations.*

Cont, R., and J. Fonseca. 2002. Dynamics of implied volatility surfaces. *Quantitative Finance* 2 (1): 45–60.

Cooperation, I., C. Schittenkopf, et al. 1999. Non-linear versus non-gaussian volatility models.

Cottrell, M., B. Girard, et al. 1995. Neural modeling for time series: a statistical stepwise method for weight elimination. *IEEE Transactions on Neural Networks* 6 (6): 1355–1364.

Curry, B., and P. Morgan, et al. 1997. Neural networks and forecasting: Orthodox' methods and new research.

Cybenko, G. 1989. Approximation by superpositions of a sigmoidal function. *Mathematics of Control, Signals, and Systems (MCSS)* 2 (4): 303–314.

Day, T.E., and C.M. Lewis. 1992. Stock market volatility and the informational content of stock index options. *Journal of Econometrics* 52 (1).

De Giovanni, D., S. Ortobelli, et al. 2008. Delta hedging strategies comparison. *European Journal of Operational Research* 185 (3): 1615–1631.

De Groot, C., and D. Würtz. 1991. Analysis of univariate time series with connectionist nets: A case study of two classical examples. *Neurocomputing* 3 (4): 177–192.

Diebold, F.X. 1988. *Empirical modeling of exchange rate dynamics*. Berlin: Springer.

Diebold, F.X., and P. Pauly. 1987. Structural change and the combination of forecasts. *Journal of Forecasting* 6 (1): 21–40.

Donaldson, R.G., and M. Kamstra. 1996. Forecast combining with neural networks. *Journal of Forecasting* 15 (1): 49–61.

Dotsis, G., and R.N. Markellos. 2007. The finite sample properties of the GARCH option pricing model. *Journal of Futures Markets* 27 (6): 599–615.

Dowd, K. 1999. *Beyond value at risk: the new science of risk management*. New York: Wiley.

Drost, F.C., and T.E. Nijman. 1993. Temporal aggregation of GARCH processes. *Econometrica: Journal of the Econometric Society* 61 (4): 909–927.

Duan, J.C. 1995. The GARCH option pricing model. *Mathematical Finance* 5 (1): 13–32.

Duan, J.C. 1996. Craking the smile. *Risk* 9: 55–59.

Duan, J.C., G. Gauthier, et al. 2006. Approximating the GJR-GARCH and EGARCH option pricing models analytically. *Journal of Computational Finance* 9 (3): 41.

Duan, J.C., G. Gauthier, et al. 1999. An analytical approximation for the GARCH option pricing model. *Journal of Computational Finance* 2.

Duan, J.C., P. Ritchken, et al. 2006. Jump starting GARCH: Pricing and hedging options with jumps in returns and volatilities. *Working Paper*.

Duan, J.C., P.H. Ritchken, et al. 2006. Approximating GARCH-jump models, jump-diffusion processes, and option pricing. *Mathematical Finance* 16 (1): 21–52.

Duan, J.C., and J.G. Simonato. 1998. Empirical martingale simulation for asset prices. *Management Science* 44 (9): 1218–1233.

Dumas, B., J. Fleming, et al. 1998. Implied volatility functions: Empirical tests. *Journal of Finance* 53 (6): 2059–2106.

Dunis, C.L., and X. Huang. 2002. Forecasting and trading currency volatility: An application of recurrent neural regression and model combination. *Journal of Forecasting* 21 (5): 317–354.

Dunis, C.L., J. Laws, et al. 2010. Modelling commodity value at risk with higher order neural networks *Applied Financial Economics* 20 (7): 585–600.

Engle, R. 2001. GARCH 101: The use of ARCH/GARCH models in applied econometrics. *Journal of Economic Perspectives* 15 (4): 157–168.

Engle, R.F. 1982. Autoregressive conditional heteroscedasticity with estimates of the variance of United Kingdom inflation. *Econometrica* 50 (4): 987–1007.

Engle, R.F., T. Ito, et al. 1990. Meteor showers or heat waves? Heteroskedastic intra-daily volatility in the foreign exchange market. *Econometrica* 58 (3): 525–542.

Engle, R.F., and C. Mustafa. 1992. Implied ARCH models from options prices. *Journal of Econometrics* 52 (1): 2.

French, K.R., G.W. Schwert, et al. 1987. Expected stock returns and volatility. *Journal of Financial Economics* 19: 3–29.

Gallant, A.R., L.P. Hansen, et al. 1990. Using conditional moments of asset payoffs to infer the volatility of intertemporal marginal rates of substitution. *Journal of Econometrics* 45 (1–2): 141–179.

Gallant, A.R., D.A. Hsieh, et al. 1991. On fitting a recalcitrant series: The pound/dollar exchange rate, 1974–1983.

Galliers, R. 1992. *Information systems research: Issues, methods, and practical guidelines.* Blackwell Scientific.

Garcia, R., and Cirano. 1998. Pricing and hedging derivative securities with neural networks and a homogeneity hint. CIRANO.

Garcia, R., and R. Gençay. 2000. Pricing and hedging derivative securities with neural networks and a homogeneity hint. *Journal of Econometrics* 94 (1–2): 93–115.

Geyer, A., and W. Schwaiger. 2001. Delta hedging with stochastic volatility in discrete time. *Financial Markets and Portfolio Management* 15 (1): 94–103.

Gibson, G.J., and C.F.N. Cowan. 1990. On the decision regions of multilayer perceptrons. *Proceedings of the IEEE* 78 (10): 1590–1594.

Gorr, W.L., D. Nagin, et al. 1994. Comparative study of artificial neural network and statistical models for predicting student grade point averages. *International Journal of Forecasting* 10 (1): 17–34.

Granger, C.W.J., and R. Ramanathan. 1984. Improved methods of combining forecasts. *Journal of Forecasting* 3 (2): 197–204.

Granger, C.W.J., and T. Teräsvirta. 1993. *Modelling nonlinear economic relationships.* USA: Oxford University Press.

Grossberg, S. 1973. Contour enhancement, short term memory, and constancies in reverberating neural networks. *Studies in applied Mathematics* 52 (3): 213–257.

Grudnitski, G., and L. Osburn. 1993. Forecasting S&P and gold futures prices: An application of neural networks. *Journal of Futures Markets* 13 (6): 631–643.

Gultekin, N.B., R.J. Rogalski, et al. 1982. Option pricing model estimates: some empirical results. *Financial Management* 11 (1): 58–69.

Hamid, S.A., and Z. Iqbal. 2004. Using neural networks for forecasting volatility of S&P 500 index futures prices. *Journal of Business Research* 57 (10): 1116–1125.

Hammer, J.A. 1989. On biases reported in studies of the black-scholes option pricing model. *Journal of Economics and Business* 41 (2): 153–169.

Hanke, M. 1997. Neural network approximation of option pricing formulas for analytically intractable option pricing models. *Journal of Computational Intelligence in Finance* 5 (5): 20–27.

Hanke, M. 1999. Neural networks versus Black-Scholes: An empirical comparison of the pricing accuracy of two fundamentally different option pricing methods. *Journal of Computational Finance* 5: 26–34.

Haykin, S. 2008. *Neural networks: A comprehensive foundation.* Prentice Hall.

Henderson, V. 2004. Black-Scholes model. *Encyclopedia of Actuarial Science* 1: 175–184.

Heston, S.L., and S. Nandi. 2000. A closed-form GARCH option valuation model. *Review of Financial Studies* 13 (3): 585.

Heston, S. L., S. Nandi, et al. 1997. *A closed-form GARCH option pricing model.* Federal Reserve Bank of Atlanta.

Hinich, M.J. 1982. Testing for Gaussianity and linearity of a stationary time series. *Journal of time series analysis* 3 (3): 169–176.

Hornik, K. 1991. Approximation capabilities of multilayer feedforward networks. *Neural Networks* 4 (2): 251–257.

Hornik, K. 1993. Some new results on neural network approximation. *Neural Networks* 6 (8): 1069–1072.

Hornik, K., M. Stinchcombe, et al. 1989. Multilayer feedforward networks are universal approximators. *Neural Networks* 2 (5): 359–366.

Hornik, K., M. Stinchcombe, et al. 1990. Universal approximation of an unknown mapping and its derivatives using multilayer feedforward networks. *Neural Networks* 3 (5): 551–560.

Hsieh, D.A. 1988. The statistical properties of daily foreign exchange rates: 1974–1983. *Journal of International Economics* 24 (1–2): 129–145.

Hsieh, K.C., and P. Ritchken. 2005. An empirical comparison of GARCH option pricing models. *Review of Derivatives Research* 8 (3): 129–150.

Hsing, S.-P. 2006. Comparison of hedging option positions of the GARCH(1,1) and the Black-Scholes models. *Master's Thesis,* National Sun Yat-Sen University Kaohsiung, Taiwan, 804, R.O.C.

Hu, M.Y., and C. Tsoukalas. 1999. Combining conditional volatility forecasts using neural networks: An application to the EMS exchange rates. *Journal of International Financial Markets, Institutions and Money* 9 (4): 407–422.

Hull, J. 2003. *Options, futures, and other derivatives.* NJ: Prentice Hall Upper Saddle River.

Hull, J., and A. White. 1987. The pricing of options on assets with stochastic volatilities. *Journal of Finance* 42 (2): 281–300.

Hung, M.S., and J.W. Denton. 1993. Training neural networks with the GRG2 nonlinear optimizer. *European Journal of Operational Research* 69 (1): 83–91.

Hutchinson, J.M., A.W. Lo, et al. 1994. A nonparametric approach to pricing and hedging derivative securities via learning networks. *The Journal of Finance* 49 (3): 851–889.

Irie, B., and S. Miyake. 1998. Capacity of three layered preceptrons. *IEEE ICNN* 1: 641–648.

Jacobs, R.A. 1988. Increased rates of convergence through learning rate adaptation. *Neural Networks* 1 (4): 295–307.

Jhee, W.C., and J.K. Lee. 1993. Performance of neural networks in managerial forecasting. *International Journal of Intelligent Systems in Accounting and Finance Management* 2 (1): 71.

Jiang, G.J., and Y.S. Tian. 2005. The model-free implied volatility and its information content. *Review of Financial Studies* 18 (4): 1305.

Jorion, P. 1988. On jump processes in the foreign exchange and stock markets. *Review of Financial Studies* 427–445.

Jorion, P. 2001. *Value at risk: The new benchmark for managing financial risk.* McGraw-Hill.

Kaastra, I., and M. Boyd. 1996. Designing a neural network for forecasting financial and economic time series. *Neurocomputing* 10 (3): 215–236.

Kang, S.Y. 1992. An investigation of the use of feedforward neural networks for forecasting. *Ph. D. Dissertation,* Kent State, 1991.

Karaali, O., W. Edelberg, et al. 1997. Modelling volatility derivatives using neural networks. *Computational Intelligence for Financial Engineering (CIFEr), 1997, Proceedings of the IEEE/IAFE 1997,* 280–286.

Klimasauskas, C.C. 1991. Neural nets tell why. *Dr. Dobb's Journal* 16 (4): 16.

Lachtermacher, G., and J.D. Fuller. 1995. Back propagation in time-series forecasting. *Journal of Forecasting* 14 (4): 381–393.

Lajbcygier, P., C. Boek, et al. 1995. Neural network pricing of all ordinaries SPI options on futures. *REFENES et al. Neural Networks in Financial Engineering. Proceedings of 3rd International Conference On neural Networks in the Capital Markets,* 64–77.

Lajbcygier, P., A. Flitman, et al. 1997. The pricing and trading of options using a hybrid neural network model with historical volatility. *Neurovest Journal* 5 (1): 27–41.

Lajbcygier, P.R., and J.T. Connor. 1997. Improved option pricing using artificial neural networks and bootstrap methods. *International Journal of Neural Systems* 8 (4): 457–471.

Lamoureux, C.G., and W.D. Lastrapes. 1990. Persistence in variance, structural change, and the GARCH model. *Journal of Business & Economic Statistics* 225–234.

Lawrence, S., and C.L. Giles. 2000. Overfitting and neural networks: Conjugate gradient and backpropagation. *Proceedings of the IEEE International Conference on Neural Networks* 114–119.

Lee, K.Y. 1991. Are the GARCH models best in out-of-sample performance? *Economics Letters* 37 (3): 305–308.

Lehar, A., M. Scheicher, et al. 2002. GARCH vs. stochastic volatility: Option pricing and risk management. *Journal of Banking and Finance* 26 (2–3): 323–345.

Lehnert, T. 2003. Explaining smiles: GARCH option pricing with conditional leptokurtosis and skewness. *Journal of Derivatives* 10 (3): 27–39.

Leland, H.E. 1985. Option pricing and replication with transactions costs. *The Journal of Finance* 40 (5): 1283–1301.

Lenard, M.J., P. Alam, et al. 1995. The application of neural networks and a qualitative response model to the auditor's going concern uncertainty decision. *Decision Sciences* 26 (2): 209–227.

Linsmeier, T.J., and N.D. Pearson. 2000. Value at risk. *Financial Analysts Journal* 56 (2): 47–67.

Lippmann, R.P. 1987. An introduction to computing with neural nets. *ARIEL* 209 (115): 245.

Locarek-Junge, H., and R. Prinzler. 1998. Estimating value-at-risk using neural networks. *Application of machine learning and data mining in finance, ECML.*

Loether, D.G.M.a.H.J. 1999. *Social research.* Allyn & Bacon.

Lopez, J.A. 2001. Evaluating the predictive accuracy of volatility models. *Journal of Forecasting* 20 (2): 87–109.

Luukkonen, R., P. Saikkonen, et al. 1988. Testing linearity against smooth transition autoregressive models. *Biometrika* 75 (3): 491.

Macbeth, J.D., and L.J. Merville. 1979. An empirical examination of the Black-Scholes call option pricing model. *Journal of Finance* 34 (5): 1173–1186.

Manfredo, M.R., and R.M. Leuthold. 2001. Market risk and the cattle feeding margin: An application of value-at-risk. *Agribusiness* 17 (3): 333–353.

McCurdy, T.H., and I.G. Morgan. 1988. Testing the martingale hypothesis in deutsche mark futures with models specifying the form of heteroscedasticity. *Journal of Applied Econometrics* 3 (3): 187–202.

McMillan, D., A. Speight, et al. 2000. Forecasting UK stock market volatility. *Applied Financial Economics* 10 (4): 435–448.

Medsker, L., E. Turban, et al. 1993. Neural network fundamentals for financial analysts. *The Journal of Investing* 2 (1): 59–68.

Meissner, G., and N. Kawano. 2001. Capturing the volatility smile of options on high-tech stocks—a combined GARCH-neural network approach. *Journal of Economics and Finance* 25 (3): 276–292.

Mello, A.S., and H.J. Neuhaus. 1998. A portfolio approach to risk reduction in discretely rebalanced option hedges. *Management Science* 44 (7): 921–934.

Merton, R.C. 1973. Theory of rational option pricing. *The Bell Journal of Economics and Management Science* 4 (1): 141–183.

Merton, R.C. 1976. Option pricing when underlying stock returns are discontinuous. *Journal of Financial Economics* 3 (1–2): 125–144.

Miazhynskaia, T., G. Dorffner, et al. 2003. Risk management application of the recurrent mixture density network models. *Artificial Neural Networks and Neural Information Processing—ICANN/ICONIP 2003,* 180–180.

Michel Crouhy, D. G., and Robert Mark (2001). Risk management.

Milhøj, A. 1987. A conditional variance model for daily deviations of an exchange rate. *Journal of Business & Economic Statistics* 99–103.

Minsky, M.L., and S. Papert. 1988. *Perceptrons: An introduction to computational geometry.* Mass: MIT press Cambridge.

Minsky, M.L., and eS Papert. 1969. *Perceptrons.* Cambridge (Massachusetts): The MIT Press.

Mitchell, T.M. 1997. *Machine learning,* 82. Burr Ridge, IL: McGraw Hill.

Morgan, J.P. 1997. *Creditmetrics-technical document.* New York: JP Morgan.

Murata, N. 1994. Network information criterion-determining the number of hidden units for an artificial neural network model. *IEEE Transactions on Neural Networks* 5 (6): 865.

Nabney, I.T., and H.W. Cheng. 1997. *Estimating conditional volatility with neural networks*, Citeseer.

Nam, K., and T. Schaefer. 1995. Forecasting international airline passenger traffic using neural networks. *Logistics and Transportation Review* 31 (3): 239–252.

Nelson, D.B. 1991. Conditional heteroskedasticity in asset returns: a new approach. *Econometrica* 59 (2): 347–370.

Ng, H.S.R. and K.P. Lam. 2006. How does sample size affect garch models? In *Proceedings Joint Conference on Information Science*.

Nunamaker Jr., J.F., M. Chen, et al. 1991. Systems development in information systems research. *Journal of Management Information Systems* 7 (3): 89–106.

Nwogugu, M. 2006. Further critique of GARCH/ARMA/VAR/EVT stochastic-volatility models and related approaches. *Applied Mathematics and Computation* 182 (2): 1735–1748.

Ormoneit, D., and R. Neuneier. 1996. Experiments in predicting the German stock index DAX with density estimating neural networks.

Peter, C., and J. Kris. 2004. Which GARCH model for option valuation? *Management Science* 50 (9): 1204–1221.

Poon, S.H. 2005. *A practical guide to forecasting financial market volatility*. New York: Wiley.

Poterba, J.M., and L.H. Summers. 1986. The persistence of volatility and stock market fluctuations. *The American Economic Review* 76 (5): 1142–1151.

Rabemananjara, R., and J.M. Zakoïan. 1993. Threshold ARCH models and asymmetries in volatility. *Journal of Applied Econometrics* 8 (1): 31–49.

Reed, R. 1993. Pruning algorithms—A survey. *IEEE Transactions on Neural Networks* 4 (5): 740–747.

Refenes, A.N. 1995. Neural network design considerations In *Neural networks in the capital market*, ed. Refenes, A. N. New York: John Wiley.

Ritchken, P., and R. Trevor. 1999. Pricing options under generalized GARCH and stochastic volatility processes. *The Journal of Finance* 54 (1): 377–402.

Roy, A., S. Govil, et al. 1995. An algorithm to generate radial basis function (RBF)-like nets for classification problems. *Neural Networks* 8 (2): 179–201.

Rubinstein, M. 1976. The valuation of uncertain income streams and the pricing of options. *Bell Journal of Economics Management Science* 7: 407–425.

Rubinstein, M. 1985. Nonparametric tests of alternative option pricing models using all reported trades and quotes on the 30 most active CBOE option classes from August 23, 1976 through August 31, 1978. *The Journal of Finance* 40 (2): 455–480.

Rumelhart, D.E., J.L. McClelland, et al. 1986. Parallel distributed processing.

Samur, Z., and G.T. Temur. 2009. The use of artificial neural network in option pricing: the case of S&P 100 index options. *Engineering and Technology* 54: 326–331.

Satchell, S., and A. Timmermann. 1995. An assessment of the economic value of non linear foreign exchange rate forecasts. *Journal of Forecasting* 14 (6): 477–497.

Schittenkopf, C., and G. Dorffner. 2001. Risk-neutral density extraction from option prices: Improved pricing with mixture density networks. *IEEE Transactions on Neural Networks* 12 (4): 716–725.

Schittenkopf, C., G. Dorffner, et al. 2000. Forecasting time-dependent conditional densities: A semi non-parametric neural network approach. *Journal of Forecasting* 19 (4): 355–374.

Schittenkopf, C., G. Dorner, et al. 1998. Volatility prediction with mixture density networks. In *Proceedings of the International Conference on Artificial Neural Networks*, accepted, Skovde, Sweden.

Schöneburg, E. 1990. Stock price prediction using neural networks: A project report. *Neurocomputing* 2 (1): 17–27.

Schwert, G.W. 1990. Stock volatility and the crash of '87. *Review of Financial Studies* 3 (1): 77–102.

Sentana, E. 1995. Quadratic ARCH models. *The Review of Economic Studies* 62 (4): 639–661.

Sharda, R. 1994. Neural networks for the MS/OR analyst: An application bibliography. *Interfaces* 24 (2): 116–130.

Sharda, R., and R.B. Patil. 1992. Connectionist approach to time series prediction: An empirical test. *Journal of Intelligent Manufacturing* 3 (5): 317–323.

Sietsma, J., and R.J.F. Dow. 1991. Creating artificial neural networks that generalize. *Neural Networks* 4 (1): 67–79.

Srinivasan, V., P. Bhatia, et al. 1994. Edge detection using a neural network. *Pattern Recognition* 27 (12): 1653–1662.

Sterk, W. 1982. Tests of two models for valuing call options on stocks with dividends. *Journal of Finance* 37 (5): 1229–1237.

Stinchcombe, M., and H. White. 1992. Using feedforward networks to distinguish multivariate populations. In *International Joint Conference on Neural Networks, IJCNN, 1992*.

Tang, Z., and P.A. Fishwick. 1993. Feedforward neural nets as models for time series forecasting. *ORSA Journal on Computing* 5: 374.

Thomaidis, N. S., and G. D. Dounias. 2006. A general class of neural network-GARCH models for financial time series analysis. *SSRN eLibrary*.

Tino, P., C. Schittenkopf, et al. 2001. Financial volatility trading using recurrent neural networks. *IEEE Transactions on Neural Networks* 12 (4): 865–874.

Tong, H. 1990. Non-linear time series. A dynamical system approach.

Tong, H., and K. S. Lim. 1980. Threshold autoregression, limit cycles and cyclical data. *Journal of the Royal Statistical Society. Series B (Methodological)* 245–292.

Tsay, R.S. 2005. *Analysis of financial time series*. Wiley-Interscience.

Vähämaa, S. 2004. Delta hedging with the smile. *Financial Markets and Portfolio Management* 18 (3): 241–255.

Vishwakarma, K.P. 1994. Recognizing business cycle turning points by means of a neural network. *Computational Economics* 7 (3): 175–185.

Wang, Z., C. Di Massimo, et al. 1994. A procedure for determining the topology of multilayer feedforward neural networks. *Neural Networks* 7 (2): 291–300.

Wasserman, P.D. 1989. *Neural computing: Theory and practice*. New York, NY, USA: Van Nostrand Reinhold Co.

Weigend, A.S., B. Huberman, et al. 1990. Predicting the future: A connectionist approach. *International Journal of Neural Systems* 1 (3): 193–209.

Weigend, A.S., and D.A. Nix. 1994. *Predictions with confidence intervals (local error bars)*. Colorado University at Boulder Department of Computer Science.

Whaley, R.E. 1982. Valuation of American call options on dividend-paying stocks: Empirical tests. *Journal of Financial Economics* 10 (1): 29–58.

Widrow, B., and S.D. Stearns. 1985. *Adaptive signal processing*, Vol. 491, 1. Englewood Cliffs, NJ, Prentice-Hall, Inc.

Wong, B.K., V.S. Lai, et al. 2000. A bibliography of neural network business applications research: 1994–1998. *Computers and Operations Research* 27 (11): 1045–1076.

Wong, F.S. 1991. Time series forecasting using backpropagation neural networks. *Neurocomputing* 2 (4): 147–159.

Wu, X., Y. Sun, et al. 2005. A quantile-data mapping model for value-at-risk based on BP and support vector regression. *Internet and Network Economics* 1094–1102.

Xiao, W., Q. Fei, et al. 2008. Forecasting Chinese stock markets volatility based on neural network combining.

Yao, J., Y. Li, et al. 2000. Option price forecasting using neural networks. *Omega* 28 (4): 455–466.

Zaiyong Tang, C. 1991. Time series forecasting using neural networks vs. Box-Jenkins methodology. *Simulation* 57 (5): 303.

Zhang, G., and B. Eddy Patuwo. 1998. Forecasting with artificial neural networks: The state of the art. *International journal of forecasting* 14 (1): 35–62.

Zhang, G., B.E. Patuwo, et al. 1998. Forecasting with artificial neural networks: The state of the art. *International Journal of Forecasting* 14: 35–62.

Zhang, W., K. Doi, et al. 1994. Computerized detection of clustered microcalcifications in digital mammograms using a shift-invariant artificial neural network. *Medical Physics* 21: 517.

Printed in the United States
By Bookmasters